Artificial Intelligence-Based State-of-Health Estimation of Lithium-Ion Batteries—2nd Edition

Artificial Intelligence-Based State-of-Health Estimation of Lithium-Ion Batteries—2nd Edition

Guest Editors

Remus Teodorescu
Xin Sui

MDPI

Basel • Beijing • Wuhan • Barcelona • Belgrade • Novi Sad • Cluj • Manchester

Guest Editors
Remus Teodorescu
Department of Energy
Aalborg University
Aalborg
Denmark

Xin Sui
Department of Energy
Aalborg University
Aalborg
Denmark

Editorial Office
MDPI AG
Grosspeteranlage 5
4052 Basel, Switzerland

This is a reprint of the Special Issue, published open access by the journal *Batteries* (ISSN 2313-0105), freely accessible at: https://www.mdpi.com/journal/batteries/special_issues/S7982CHZIN.

For citation purposes, cite each article independently as indicated on the article page online and as indicated below:

Lastname, A.A.; Lastname, B.B. Article Title. *Journal Name* **Year**, *Volume Number*, Page Range.

ISBN 978-3-7258-6187-3 (Hbk)
ISBN 978-3-7258-6188-0 (PDF)
https://doi.org/10.3390/books978-3-7258-6188-0

Contents

About the Editors

Remus Teodorescu

Remus Teodorescu (IEEE Fellow) received a Dipl.Ing. degree in electrical engineering from the Polytechnical University of Bucharest, Bucharest, Romania, in 1989, a Ph.D. degree in power electronics from the University of Galati, Romania, in 1994 and a doctorate in 2016 from Transilvania University of Brasov.

In 1998, he joined the Department of Energy Technology at Aalborg University where he is currently a Full Professor. Between 2013 and 2017, he has been a Visiting Professor with Chalmers University. He has been a IEEE/PELS Fellow since 2012 for contributions to grid converter technology for renewable energy systems. In 2022, he became a Villum investigator and leader of Center of Research for Smart Battery (CROSBAT) at Aalborg University. His main current research areas are in the following: modular multilevel converters (MMCs) for HVDC/FACTS, Li-ion battery SOH estimation with AI and smart batteries.

Xin Sui

Xin Sui (Senior Member, IEEE) received a B.Eng. degree from Northeast Electric Power University, Jilin, China, in 2015, and an M.Sc. degree in from Institute of Electrical Engineering, Chinese Academy of Sciences, Beijing, China, in 2018, both in electrical engineering. In 2022, she received a Ph.D. degree in machine learning for battery state of health estimation from Aalborg University (AAU), Aalborg, Denmark.

She was a visiting researcher with the Department of Electrical Engineering, Chalmers University of Technology, Gothenburg, in 2023, with the Department of Mechanical Engineering, Imperial College London, London, UK, in 2024, and with the SINTEF Digital and Norwegian Research Center for AI Innovation, Oslo, Norway, in 2025.

She was an Assistant Professor in the Department of Energy, Aalborg University (AAU), Denmark, and is currently a Senior Research Scientist at the Institute of Energy, Hefei Comprehensive National Science Center, China. Her research interests include battery modeling, battery states estimation, lifetime prediction, feature engineering and machine learning.

Dr. Sui was identified as being among the world's top 2% scientists by Stanford University in 2024 and 2025. She received the Nordic Energy Challenge Award in 2025 and MSCA Seal of Excellence award in 2023. She serves as an Early Career Editorial Board Member of the journal *Batteries* as well as a Guest Editor of featured Special Issue on Artificial Intelligence-Based State-of-Health Estimation of Lithium-Ion Batteries. She is a Technical Program Committee member for IET Energy Storage Track at Powering Net Zero Week and a member of the Editorial Board of *Frontiers in Energy Research*, as well as having served as a session chair at IEEE conferences multiple times.

Preface

The rapid growth of electrified transportation and renewable energy systems puts high demands on the performance, safety, and lifetime of lithium-ion batteries. Accurate State-of-Health (SOH) estimation is becoming increasingly significant for ensuring reliable operation and enabling advanced battery management strategies. In recent years, Artificial Intelligence (AI) has emerged as a powerful tool to complement conventional electrochemical and model-based approaches, offering improved capability to capture nonlinear degradation behavior and to extract meaningful aging features from diverse and large operational datasets.

This Reprint gathers contributions from leading researchers working on data-driven, physics-informed, and hybrid AI techniques for battery SOH estimation. Specifically, the collected works cover feature extraction, degradation modelling, real-time SOH estimation, and early ageing warning, reflecting the development of AI-enabled battery informatics and analytics. We hope that this Reprint provides readers with a clear overview of current progress, highlights emerging challenges, and inspires future research toward more reliable and interpretable SOH estimation algorithms.

Remus Teodorescu and Xin Sui
Guest Editors

batteries

MDPI

Article

Design Optimisation of Metastructure Configuration for Lithium-Ion Battery Protection Using Machine Learning Methodology

Indira Cahyani Fatiha [1], Sigit Puji Santosa [1,*], Djarot Widagdo [1] and Arief Nur Pratomo [2]

[1] Mechanics of Solids Lightweight Structures Research Group, Faculty of Mechanical and Aerospace Engineering, Institut Teknologi Bandung (ITB), Jalan Ganesha 10, Bandung 40132, Indonesia
[2] Department of Mechanical Engineering, Faculty of Defense Science and Technology, Indonesia Defense University, Indonesian Peace and Security Centre (IPSC), Bogor 16810, Indonesia
* Correspondence: sigit.santosa@itb.ac.id

Abstract: The market for electric vehicles (EVs) has been growing in popularity, and by 2027, it is predicted that the market valuation will reach $869 billion. To support the growth of EVs in public road safety, advances in battery safety research for EV application should achieve low-cost, lightweight, and high safety protection. In this research, the development of a lightweight, crashworthy battery protection system using an excellent energy absorption capability is carried out. The lightweight structure was developed by using metastructure constructions with an arrangement of repeated lattice cellular structures. Three metastructure configurations (bi-stable, star-shaped, double-U) with their geometrical variables (thickness, inner spacing, cell stack) and material types (stainless steel, aluminium, and carbon steel) were evaluated until the maximum Specific Energy Absorptions (SEA) value was attained. The Finite Element Method (FEM) is utilised to simulate the mechanics of impact and calculate the optimum SEA of the various designs using machine learning methodology. Latin Hypercube Sampling (LHS) was used to derive the design variation by dividing the variables into 100 samples. The machine learning optimisation method utilises the Artificial Neural Networks (ANN) and Non-dominated Sorting Genetic Algorithm-II (NSGA-II) to forecast the design that produces maximum SEA. The optimum control variables are star-shaped cells consisting of one vertical unit cell using aluminium material with a cross-section thickness of 2.9 mm. The optimum design increased the SEA by 5577% compared to the baseline design. The accuracy of the machine learning prediction is also verified using numerical simulation with a 2.83% error. Four different sandwich structure configurations are then constructed using the optimal geometry for prismatic battery protection subjected to ground impact loading conditions. An optimum configuration of $6 \times 4 \times 1$ core cells arrangement results in a maximum displacement of 7.33 mm for the prismatic battery in the ground impact simulation, which is still less than the deformation threshold for prismatic battery safety of 10.423 mm. It is shown that the lightweight metastructure is very efficient for prismatic battery protection subjected to ground impact loading conditions.

Keywords: battery protection; metastructure; machine learning; crashworthiness

1. Introduction

In recent years, electric vehicle usage and popularity have been growing steadfastly. The Mobility Market Outlook report from Statista estimates that the worldwide sales of electric vehicles will reach $384 billion by 2022 [1]. According to the analysis, in 2027, it is predicted that the combined market value of electric vehicles will be $869 billion, surpassing the sales of internal combustion engine vehicles. Global electric vehicles are too new to project a certain future trajectory, but the market is expected to accelerate exponentially due to their cost/energy efficiency, improved performance, and limited environmental impact. Over 3.2 million new plug-in vehicles were listed worldwide in the year's first half [2].

The increasing growth of electric vehicles has increased the significance of battery usages while also emphasising the need to develop their safety. Lithium-ion batteries are currently used in most of today's electric vehicles due to excellent performance at high temperatures, low self-discharge, high power-to-weight ratio, and elevated energy efficiency [3]. Research and development have proceeded to lower the cost of Li-ion batteries, increase their usable life, and solve overheating safety concerns.

While the batteries are intact and free of flaws, lithium batteries are usually secure and not likely to explode. Li-ion battery damage from a crash can happen quickly or gradually and present the risks of fire or explosion [4]. The Li-ion battery in a moving vehicle always functions at different speeds and accelerations. A moving battery is subject to local stresses and deformations that, in extreme circumstances, such as car accidents, may cause local cell damage [5]. An increase in battery temperature could set off more unfavourable mechanisms and result in thermal runaway, an uncontrollable situation where the battery produces too much heat [6]. Therefore, safeguarding Li-ion cells from harm caused by crashes is a key challenge. Battery packs are positioned away from impact areas and from where foreign things could penetrate. However, for bigger electric vehicles, the floor battery pack arrangement exposes the battery to serious damage from ground impact, including localised indentation, piercing, and fracture of its bottom construction [7].

Improving the protection system becomes a priority to limit deformation during impact and lower incident risk. A crashworthy battery protector should be extremely light to conserve energy, strong enough to prevent excessive deformation, and have excellent energy absorption capabilities. Many designs, including sandwich structures, have been researched as Li-ion battery shields. The sandwich structure consists of a pair of skins that is strong, thin, and stiff, which becomes a cover on the upper and lower parts; the thick but light core material is placed in the middle of the skins as a load transfer medium, usually using cell-shaped material with less rigid and ultra-low-density material. These structures' light weight and high energy absorption capability make them popular in the construction and aerospace industries [8,9].

Different structures, including metastructure, can be placed within the sandwich structure's core [9]. Metastructures, which are lighter than comparable solid structures, form by repeatedly arranging unit cell structures. Metastructure can be classified using a variety of structures, with human bones and honeycomb-like beehive structures as two examples found in nature. Because of its lightweight characteristics and strength, metastructure has significant potential for high energy absorption. For this reason, the metastructure-filled sandwich structure make it a viable candidate for use as a battery protection system.

One of the smallest unit cells in the metastructure is the bi-stable element, which is becoming increasingly popular as a basic element due to its ability to easily support loads and be manufactured using additive manufacturing techniques [10]. While other bi-stable unit cell components appear rigid, the curved section is first constructed as a curved clamped beam [11]. Their multi-stability allows their curved shell designs to create newly architected metamaterials to trap energy [12]. Due to the instability of the buckling that initiates the bi-stable snap-through behaviour, this structure with negative stiffness characteristics will experience greater deformation, preserve its deformed shape, and trap the crushing energy. As a result, it is very suitable as an energy-absorbing element.

Another type of metastructure is the auxetic structure [13]. Due to its exact geometric configuration and mechanical deformation of its microstructures, auxetic material shrinks laterally when crushed because it has a negative Poisson's ratio [14]. As a result, this material exhibits exceptional shear stiffness, excellent fracture toughness, remarkable indentation resistance, and unusual energy absorption capabilities. Because the auxetic structure is drawn to the impact location, it also has a smaller peak crushing force, which lessens damage and harm to the protected object [14]. With its remarkable and unique features, the auxetic structure has great potential for many applications, such as protection work and variable-type aerospace wingtips [15]. The auxetic patterns explored in this research are star-shaped and double-U honeycomb (DUH).

Previous research has been focused on the optimisation design of different metastructure configurations with a wide range of optimisation methods, including topological, multi-objective, Design for Six Sigma (DFSS), machine learning-based, etc. Filho et al. [16] performed a multi-objective optimisation of sustainable sandwich panels with perforated foam cores to withstand impact loads. The research by Francisco et al. [17] centred on optimising the double arrowhead auxetic model using the multi-objective Lichtenberg algorithm. Nasrullah et al. [18] explored the topology optimisation of lattice structure configuration design for crashworthy components. The study by Kirana [19] carried out the optimisation of the bi-stable metastructure configuration subjected to explosive and compressive loads to present the energy-absorbing mechanism using the Taguchi method with an L16′ orthogonal array. Biharta et al. [20] developed an auxetic-based 3D sandwich architecture to protect an electric vehicle pouch battery from ground impact stress. The auxetic structures were optimised using Taguchi's DFSS method to increase the specific energy absorbed. Another relevant research is from Carakapurwa et al. [21], who used machine learning to predict the best protection system for pouch batteries by optimising several 2D auxetic configurations (re-entrant, double-arrow, star-shaped, DUH), material, and geometric variables.

To address the gaps from previous research regarding application and design variables, this research will focus on optimising the electric vehicle prismatic battery protection system to resist ground impact using a sandwich metastructure configuration. Based on studies by Xia et al. [7], the sandwich structure was positioned on the floor structure of the vehicle and simulated to withstand an impact from an oncoming impactor that came from below. The prismatic Li-ion battery modelled and experimentally tested in the study by Reynolds et al. [22] formed the basis for the battery system used in this research.

Bi-stable [19], star-shape [23], and double-u honeycomb [24] 3D configurations were the different types of geometry for the metastructure studied, while the material type was limited to SS304, Al6061, and ST37. The machine learning optimisation method was implemented to vary these variables to forecast the ideal configuration. To acquire the SEA for every sample in the data sampling process, 100 samples were constructed using LHS and manually simulated through FEM. The whole input and SEA values datasets were then processed through data training using ANN to develop a mathematical model that predicted SEA. As the final optimisation stage, the NSGA-II was used to iterate the variables in search of the maximum SEA.

The optimised metastructure design predicted by machine learning was modelled in an entire battery system simulation exposed to a dynamic impact to validate its accuracy and safety capability. The crashworthiness of the entire battery system and optimised structure was evaluated using a prismatic battery failure threshold, which fails when the deformation exceeds 10.423 mm [22]. If the battery's deformation is less than the failure limit, the system can safeguard against severe deformation.

The objective of the battery protection structure's optimisation and analysis is to offer the best possible solution for the safety of electric vehicle Li-ion batteries. Future uses for electric vehicles may benefit from implementing the design of battery protection against thermal runaways presented in this research. Machine learning optimisation can also offer strong solutions to determine outstanding SEA value.

2. Geometry and Numerical Model

2.1. Bi-Stable Design

Optimisation of 3D bi-stable meta-structure using Taguchi's DFSS method had been researched for explosive and compressive loads. Basic geometry from Kirana's optimised 3D beam-type metastructure [19] was chosen as the preliminary baseline model to study its crashworthiness performance. The geometric model replicated the model from Zhang et al. [25] by adjusting several parameters. Preliminary modelling and simulation were used to find accurate settings and validate the model. The dimensions listed in Table 1 Figure 1 were used to build the baseline model. Two, two-dimensional unit cells were combined and joined in the middle to form a 3D metastructure unit cell, as shown in Figure 2.

Table 1. Metastructure configuration unit cell specifications.

Parameters	Value (mm)
t	0.9
h	6
$L/2$	15
$w = H2 = H4$	2.5
wm	5
$H1$	9.5
$H3$	8.6
l	40
$Htot = H3 + H1 + h + t$	25

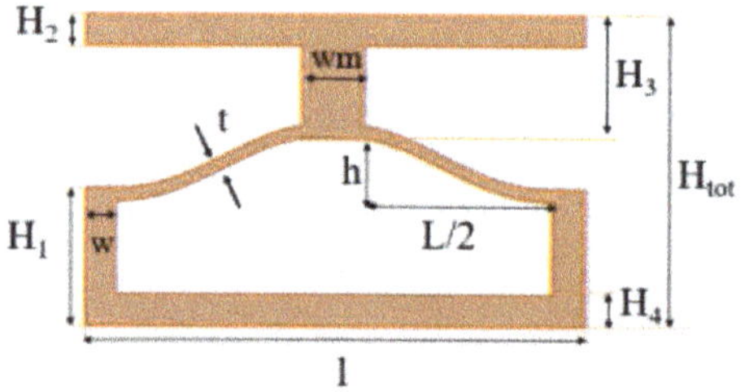

Figure 1. 2D unit cell of beam-type metastructure configuration.

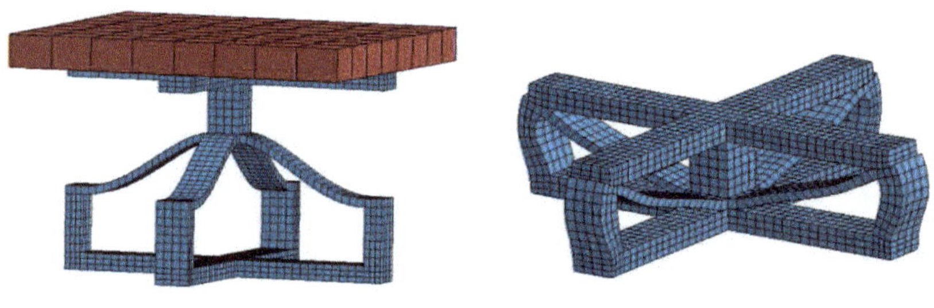

Figure 2. 3D bi-stable baseline configuration with an impactor (**left**), and its crushing result (**right**).

The simulation was performed using LS-DYNA explicit numerical simulation. The impactor is modelled as a solid rigid box that measures 30 × 30 × 5 mm, weighs 500 kg, and moves at a constant 15 m/s velocity. The material used for this model is Stainless Steel SS304 with properties defined from MatWeb, Blacksburg, VA, USA [26].

2.2. Result Validation

To validate the result of this simulation, the force vs. displacement and energy vs. displacement curves were observed in Figures 3 and 4. The slow increase in structural internal energy during quasi-static loading shows the structure's capacity to absorb impact energy. Over time, the kinetic energy value remains constant and is less than the internal energy. This is consistent with the definition of quasi-static, where the process occurs little by little and resembles a static case in which the inertial effect can be negligible. Furthermore, the sliding and hourglass energy values are both positive and significantly small in relation to the total energy. Therefore, the numerical model is presumed to be valid enough to define the actual model.

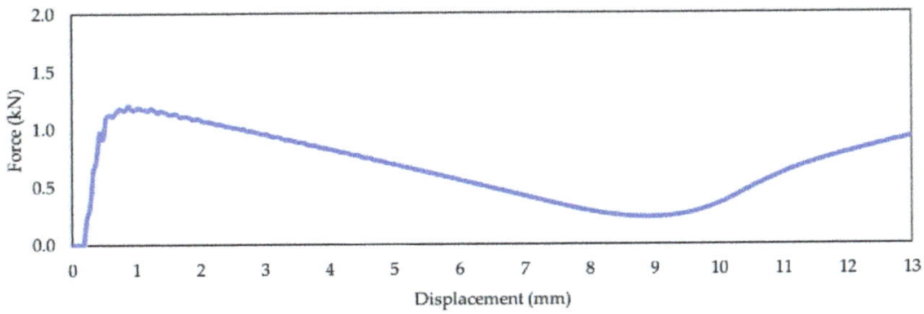

Figure 3. Crushing force vs. displacement curve baseline model.

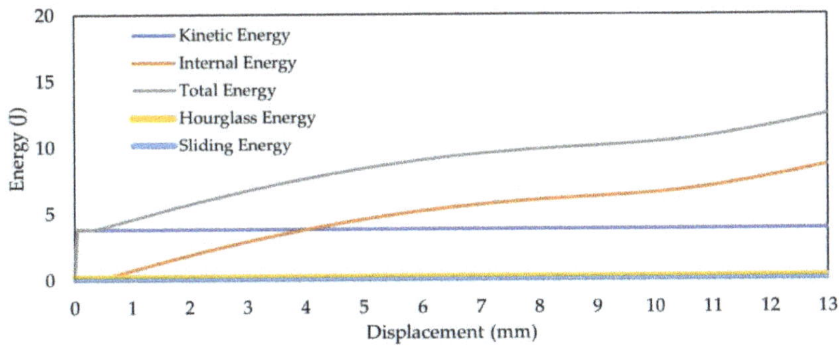

Figure 4. Energy vs. displacement curve baseline model.

3. Metastructure Configurations Modelling and Machine Learning Optimisation

3.1. Data Sampling Process

Estimating the size of the sample required for the ANN model's input was done via data sampling. The training and the validation data are two separate datasets for this data. With the help of these data, a regressor model is constructed to forecast the variables of an output from an input. Data are employed in two ways: discrete variables (categories) for cross-section configuration and material type, and continuous variables (numerical) for geometrical variables and material thickness, as described in Table 2.

Table 2. Variables for design optimisation.

Predictor Variable				**Output**
Continuous (Numerical) Variable		**Categorical (Discrete) Variable**		
Inner cell spacing	9.5–11.2 mm	Cross section shape	Bi-stable, Star, DUH	Specific Energy Absorption (SEA)
Vertical stack units	1, 2, 3			
Material thickness	1–3 mm	Material Type	SS304 [26], Al6061-T6 [27–29], ST37 [26]	

3.1.1. Geometry Definition

The geometry variables are normalized or standardized since shape and geometry properties differ. Two predetermined input variables serve as the foundation for the normalisation: h as the "inner cell spacing" and vertical stack units. The height of cells is constrained to be 29–31 mm and the configuration is set to be 2 × 2 for all samples. Variable transformation provides the geometry limitations for each cross-section type and addresses compatibility issues between input variables and the distinctive cell shapes. For example, the cell becomes wider and hollower as the inner spacing grows. The specification of the geometry constraint for each type of cross-section is listed in Table 3. The configurations of the structure are shown in Figure 5.

Table 3. Geometrical limitations for design optimisation.

Cross-Section Shape	Geometrical Limitations
Bi-stable	$h' = h \rightarrow 9.5\text{ mm} \leq h' \leq 11.2\text{ mm}$
Star-shaped	$h' = 0.834h \rightarrow 7.92\text{ mm} \leq h' \leq 9.34\text{ mm}$
DUH (Double-U Honeycomb)	$h' = 2h \rightarrow 19\text{ mm} \leq h' \leq 22.4\text{ mm}$

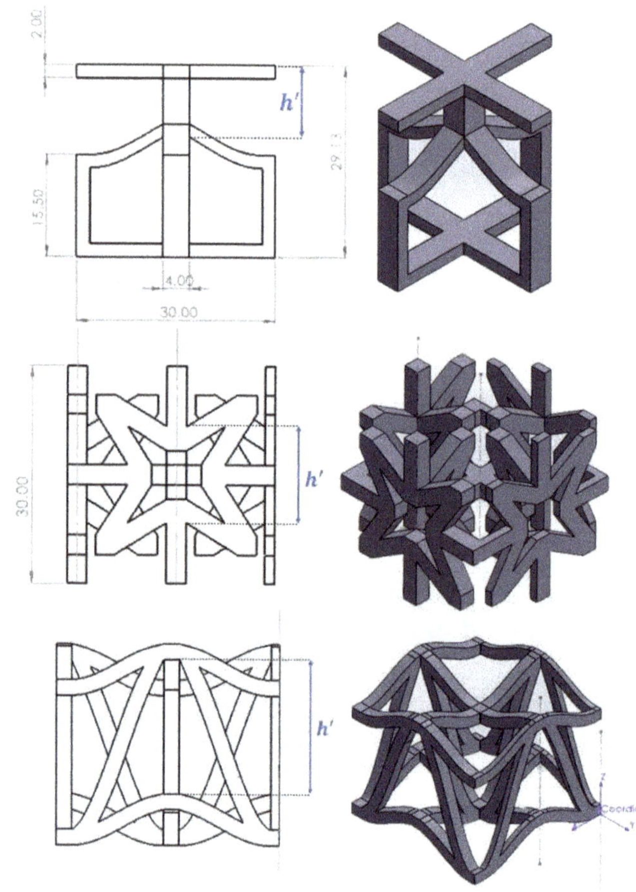

Figure 5. Variable definitions of Bi-stable, Star-shaped, and DUH cells in mm.

3.1.2. Material Definition

The material variants include stainless steel SS304, Aluminium Al 6061-T6, and carbon steel ST37. All the materials are modelled using MAT_024 piecewise linear plasticity. Al 6061-T6 properties are provided by ASM Material Data [27], Garcia's article [28], and Abotula's paper [29]. Stainless steel material is specified using material properties obtained in MatWeb [26]. The material properties are defined in Table 4.

Table 4. Material properties.

Category	Variable	SS304	Al 6061-T6	ST37	Unit
Mechanical Property	Density (ρ)	8000	2700	7330	kg/m^3
	Young's Modulus (E)	200	68.9	200	GPa
	Yield strength (σ)	290	276	290	MPa
	Poisson's ratio (v)	0.29	0.33	0.3	–
Strain-rate Sensitivity	C	100	25,000	6.844	–
	P	10	0.95	4.12	–

3.1.3. Sampling Process

The selection of samples should be broad and varied but also restricted to shorten the calculation time required to create a precise and representative model. A popular and widely utilised splitting ratio of 80:20 is employed to achieve the data-splitting approach [30]. Therefore, the sample size is 100, of which 20 will be utilised for the validation data set and 80 for the training data set. By comparing the numerical simulation results with the machine learning prediction of SEA, the error can be determined.

The discrete uniform distribution is used to create samples from categorical data, assuming that each individual has an equal probability of being chosen for a sample [31]. Meanwhile, for the continuous variables, LHS divides each parameter's range into intervals equal to the number of samples needed, then chooses one value as the midpoint from each interval to generate samples by taking into consideration the already presented sample points in a virtually random sample, which is more beneficial for multi-dimensional distribution [32,33]. As a result, it can guarantee that the range of possible values is suitably represented by the collection of random numbers.

In this study, the representative sample points are chosen randomly from 100 samples and three continuous variables to create an accurate model. Figure 6 (continuous variable), Figure 7 (categorical variable), and Figure 8 (numerical sampling space distribution) show the distributions of the sample variables created. The SEA value was then determined by numerical simulation using the sampled data as input. The data set is finished once the output value has been collected.

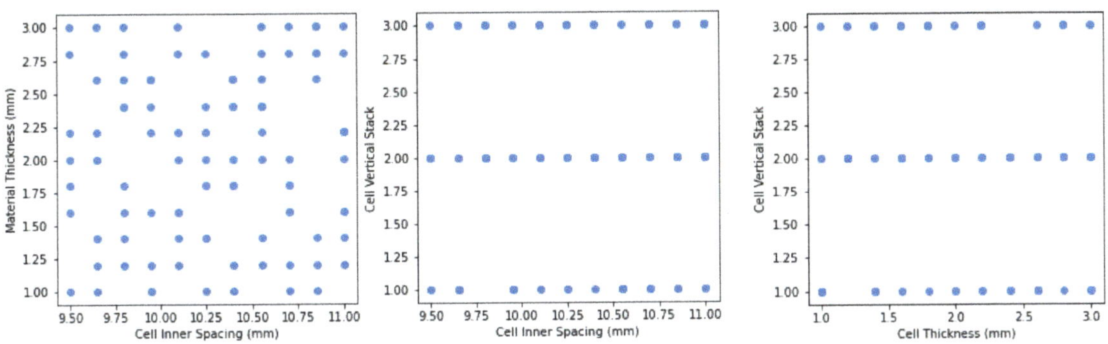

Figure 6. Continuous variable sampling result.

The samples are then modelled and imported into the LS-DYNA R11.0.0 software to determine the generated samples' crush resistance performance. The meshing process uses HyperWorks 2022 software to consider the contours of the curved structures. The samples modelled are illustrated in Figure 9.

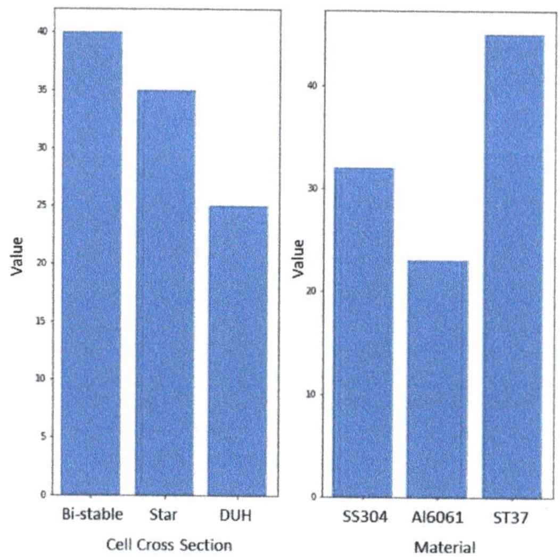

Figure 7. Categorical variable sampling result.

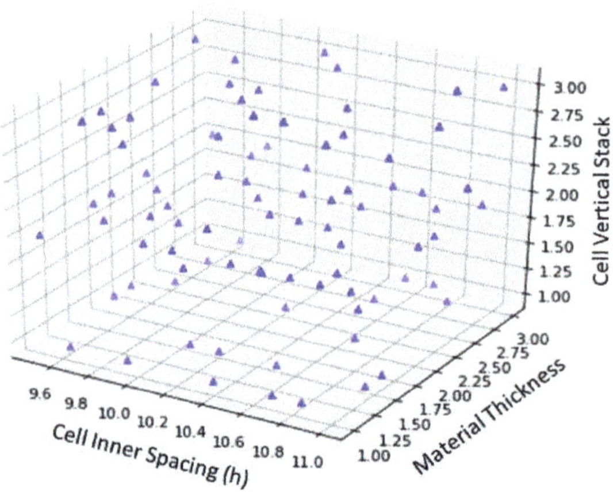

Figure 8. Continuous variable sampling result in 3D space.

The output variable is the energy absorbed per unit mass of the collision-impacted structure, as represented by SEA. The goal is to have as high an SEA as possible, given that this results in the best energy absorption capability. SEA is therefore assessed as the outcome of the optimisation.

3.2. Data Pre-Processing

The datasets presented in Table 5 are pre-processed before the input and output variables are combined. To properly quantify categorical data, they must be transformed into unique index numbers (1, 2, 3) and an artificial variable must be created to address each of these variables separately. The categorical input variables are quantified using one-hot encoding and converted into Boolean dummy models for regression analysis. The dummy variable assigned by the one-hot encoder algorithm for the input parameter's first level is erased to avoid multi-collinearity between two variables, or the dummy trap, which occurs when the variables are highly correlated. The output variables are scaled simultaneously so that the highest absolute value of each variable is 1.0 without affecting the sparsity of the data.

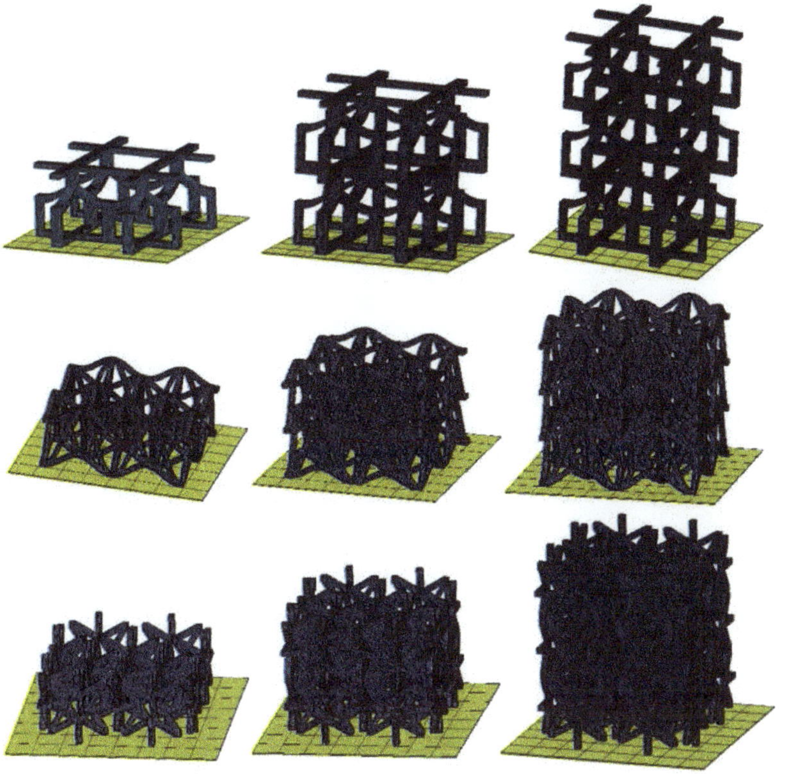

Figure 9. Numerical simulation of Bi-stable, DUH, and Star configurations.

Table 5. Categorical variables converted to dummy variables.

Categorical Variable		Number Variable	Dummy Variable		
Cross-section shape	Bistable	1	1	0	0
	DUH	2	0	1	0
	STAR	3	0	0	1
Material type	SS304	1	1	0	0
	Al6061	2	0	1	0
	ST37	3	0	0	1

In training the neural network model, the processed datasets are split into an approximate 80:20 ratio between the training and validation data to ensure accuracy and prevent overfitting [34]. The training set is instructed to understand the correlation connecting the input variables and select the best weights. The test set calculates the loss between the two datasets to assess the performance of the trained model.

3.3. ANN Architecture Model

The ANN model is used to solve optimisation questions as the regressor. To define a neural network model's confidence level, the right activation units and number of hidden layers are essential; therefore, numerous iterations are necessary to find the ideal configuration. Three hidden layers, each with a consistent 9-neuron unit count, make up the model configuration, as depicted in Figure 10. To reduce the size of the error or loss function, this number of hidden layers and units was selected. Each hidden layer's parameters are listed in Table 6. Additionally, Table 7 defines the input variables that feed into the input layer. The parameters used for the ANN optimisation are listed in Table 8.

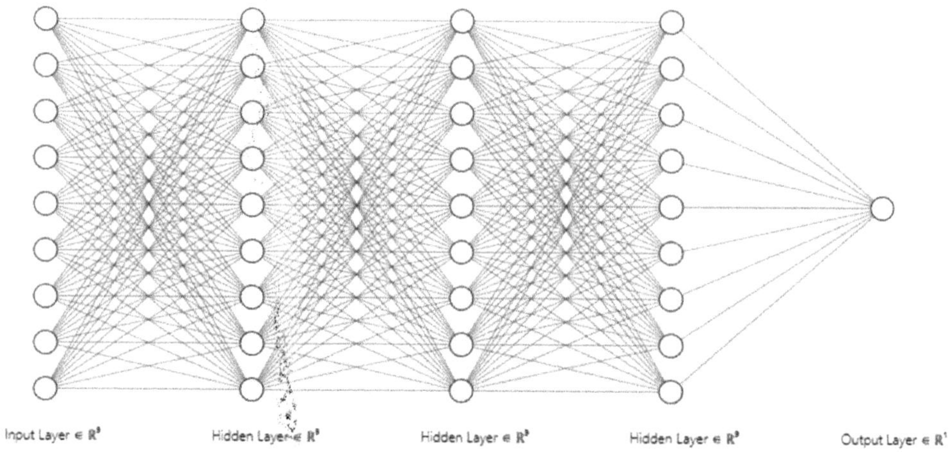

Figure 10. The architecture of ANN model.

Table 6. Units, weight, bias, and activation function for all ANN layers.

Layer	Neutron Units	Weight	Bias	Activation Function
Dense_1	9	81	9	Linear
Dense_2	9	81	9	ReLU
Dense_3	9	81	9	Sigmoid
Dense_4	9	81	9	ReLU
Dense_5 (Output: SEA)	1	9	1	Linear

Table 7. ANN input variables.

Input	Variable	Data Type
1	Dummy 2 Cell Stack (2 stack)	Boolean (0/1)
2	Dummy 3 Cell Stack (3 stack)	Boolean (0/1)
3	Dummy 2 Cross-Section (STAR)	Boolean (0/1)
4	Dummy 3 Cross-Section (DUH)	Boolean (0/1)
5	Dummy 1 Material (SS304)	Boolean (0/1)
6	Dummy 2 Material (Al6061)	Boolean (0/1)
7	Dummy 3 Material (ST37)	Boolean (0/1)
8	Inner Spacing	Float (9.8–11.3)
9	Thickness	Float (1–3)

Table 8. Variables for ANN optimisation.

Variables	Value
Epoch	3000
Initialisation	Normal distribution
Learning batch	32
Epoch step	300
Validation data split	20%
Activation Function	ReLU (Rectified Linear Unit), Sigmoid, Linear
ANN Optimizer	Adam
Error Parameter	MSE (Mean Squared Error)

3.4. ANN Model Accuracy

By comparing the numerically computed results from simulation with outcomes predicted by machine learning, the accuracy of ANN prediction can be seen in Figure 11. To determine how well the model corresponds to the real data points, we quantified the graph and obtained $R^2 = 0.9975$. The higher the value of R^2, the higher the accuracy. It is generally regarded as sufficiently precise for $R^2 > 0.9$ [35].

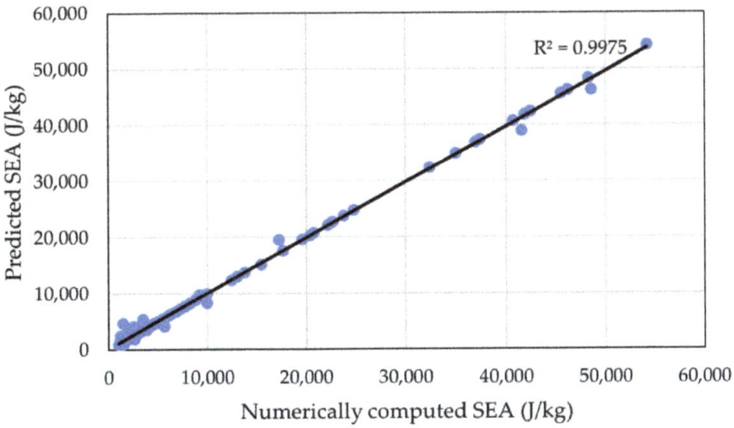

Figure 11. Comparison between predicted and computed SEA.

It is also measured for the training and validation data sets using MAE (Mean Absolute Error) and MSE, as shown in Table 9. The constructed ANN model delivers comparatively small absolute error of 0.0695 (66.95%) according to the iterations performed.

Table 9. ANN model training loss result.

Parameter	Training Set	Validation Set
MSE	2.6315×10^{-5}	0.0129
MAE	0.0013	0.0695

A neural network regression model is defined in which weight and bias values represent the connections between every layer in a neural network to comprehend the relationship between the input and output parameters. The total acquired weight obtained by the ANN model is equal to the number of layers times the number of branches in every layer, whereas the full bias attained is equal to one for every unit. Linear equations cannot be constructed from the generated weight and bias due to the Sigmoid and ReLU activation functions used in the model. Rather, the ANN model was imported straight to the NSGA-II optimisation method.

3.5. Decision of Optimised Model

The set of design parameters for the highest SEA was chosen by the NSGA-II algorithm using the trained ANN model as the input. The weight and bias of the crashworthiness variables generated from the neural network training are the first values imported by the NSGA-II algorithm. This algorithm iterated over the values of the variables according to the specified constraints and objectives. The geometrical and material definition range served as the constraints, and the objective of the NSGA-II algorithm was confined to forecasting the best crashworthiness parameters of the structure for any given arrangement of input variables.

The design variables are iterated 1000 times to produce design point candidates that meet the maximum SEA value. Based on the optimisation result, the NSGA-II algorithm produced the highest estimated SEA of 54,992.62 J/kg. Table 10 listed the design parameters of the optimised design.

Table 10. Variable maximum correlation.

Parameters	Value	Unit
Cross-section	Star	-
Material	Al6061-T6	-
Vertical stack unit	1	-
Inner spacing	10.2	mm
Material thickness	2.9	mm

Then, LS-DYNA is used to model the resulting design variables to verify the precision of ANN prediction and the result of NSGA-II optimisation (Figure 12). Table 11 compares predicted design variables, simulation output values, and baseline model.

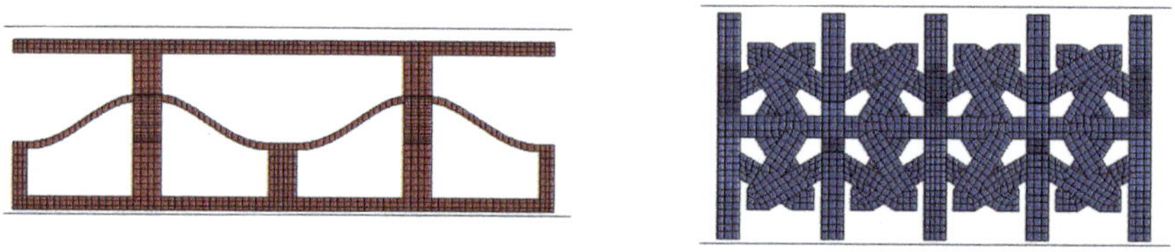

Figure 12. Numerical simulation of the baseline (**left**) and optimised (**right**) models.

Table 11. Comparison between baseline and optimised models.

Input Parameters	Baseline Model		Optimised Model	
	Value	Unit	Value	Unit
Cross-section	Bistable	-	Star	-
Material	SS304	-	Al6061	-
Vertical stack	1	-	1	-
Inner spacing (h)	14.6	mm	10.2	mm
Thickness (t)	0.9	mm	2.92	mm
Output Parameters	Value	Unit	Value	Unit
Mass	0.0589	kg	0.049	kg
EA	58.7	J	2772.38	J
SEA	996.93	J/kg	56,596.28	J/kg
Predicted SEA	-	J/kg	54,992.62	J/kg
Differences between Predicted and Simulation results				2.83%
Comparison between Optimised and Baseline SEA				5577%

As opposed to the optimised design from Kirana's research [19] as the baseline model, the optimised model's SEA value is 5577% higher. The predicted model could successfully estimate the SEA value with the difference between the predicted and actual SEA from the numerical simulation of only 2.83%. Figure 13 displays the force vs. displacement curves for the baseline and optimised models.

The optimised model's crushing force is significantly greater than the baseline model's, leading to a notable increase in energy absorption. Figure 14 also displays the energy plot for both models.

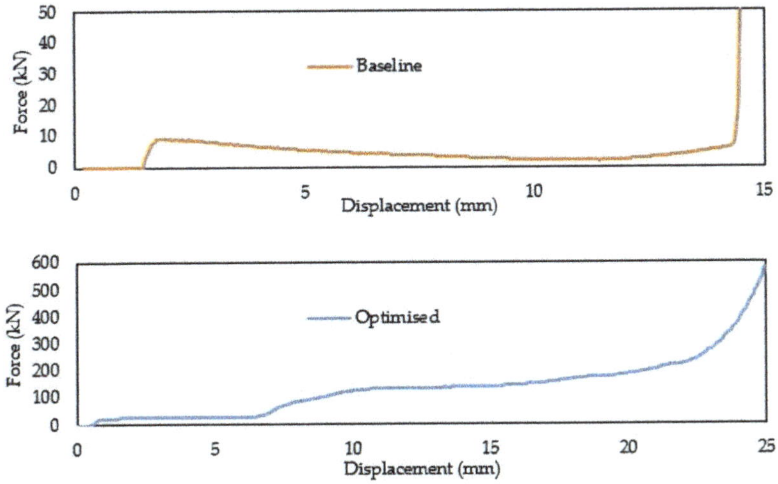

Figure 13. Force vs. displacement curves for baseline (**upper**) and optimised (**lower**) models.

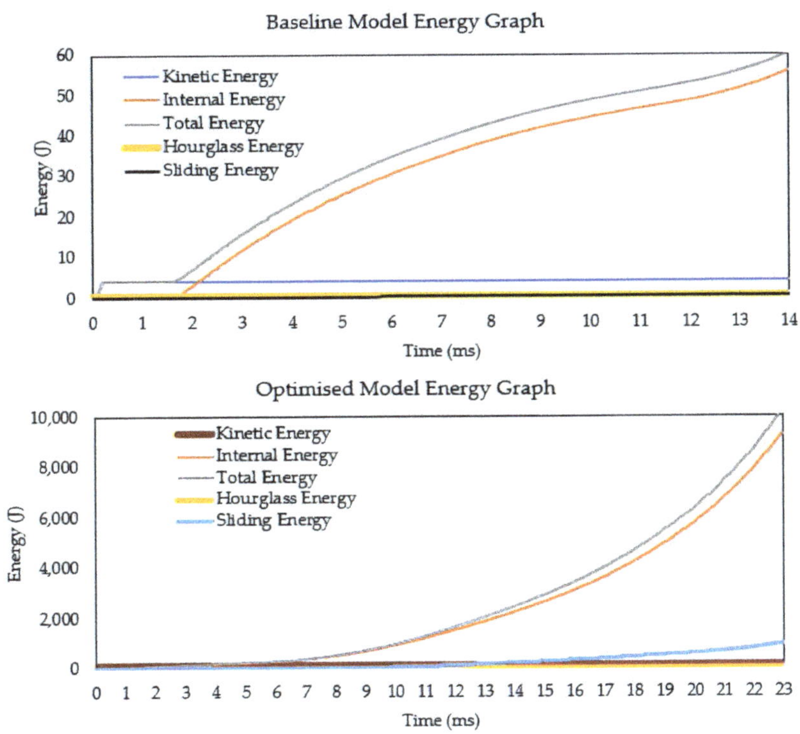

Figure 14. Energy vs. time curves for baseline (**upper**) and optimised (**lower**) models.

Notably, the impactor moves quasi-static, producing a constant kinetic energy. The energy curves exhibit a continuous rise in internal energy and a lack of negative energy, allowing the structure to absorb impact energy.

4. Battery System Analysis with Optimised Protection Configuration

4.1. Decision of Optimised Model

The final simulation shows the effectiveness of the optimised configuration in a ground impact scenario using a prismatic battery positioned on the floor structure of a vehicle. The system is based on the Lightweight Structures Laboratory ITB's model with minor adjustments from Carakapurwa's research [21]. The system comprises a prismatic battery, a floor, a plate casing for the protection layout, and an impactor, as shown in Figure 15.

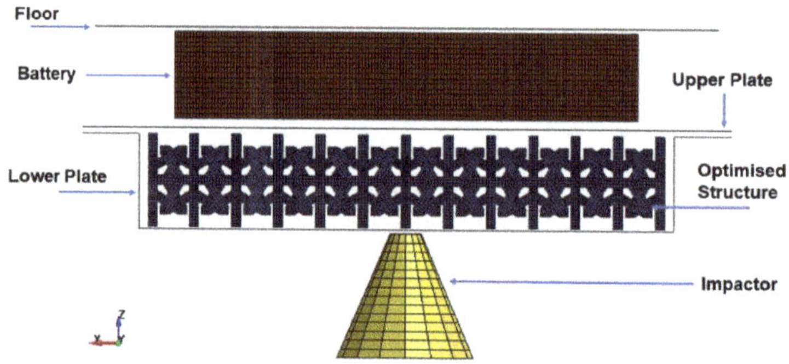

Figure 15. The complete layout of the battery system.

The optimised structure's energy absorption capability is to minimise substantial prismatic battery deformation. The failure displacement for the prismatic battery before a runaway is 10.423 mm, in light of Reynolds's study [22]. This configuration is qualified for battery protection if the structure can stop the battery's failure.

The vehicle's floor serves as its structural framework, and the plate casing is arranged in a sandwich pattern, with the cell structure serving as its core. To illustrate a foreign object that can cause damage to the battery pack, the impactor is represented as a cone. The lower and upper plates are modelled using the same material as the floor structure, as summarised in Table 12.

Table 12. Al2024-T351 material properties.

Material Properties	Value	Unit
Material used	Al 2024-T351 [36]	–
Density (ρ)	2.78×10^{-6}	kg/mm^3
Young's Modulus (E)	73.1	GPa
Yield Strength (σ)	0.324	MPa
Poisson's Ratio (v)	0.33	–
Failure Strain (ϵ_f.)	0.2	–

For the prismatic battery, all modelling parameters are based on variables and assumptions in research on prismatic battery cells. To accurately represent the homogenized behaviour of the cell under compression via thickness, material model 63 was chosen. Wei's research [37] generated the hardening curves (true stress vs. plastic strain) for the aluminium casing of prismatic battery cells, as shown in Figure 16. Based on the slope of the load-deflection curve, the modulus of elasticity is computed as a reference value for the battery module's engineering stiffness under tensile loading. Since the material is expected to be highly compressive because of the cell porosity, Poisson's ratio is set to 0.1. The dimensions of floor structure, prismatic battery, upper and lower plates are detailed in Table 13.

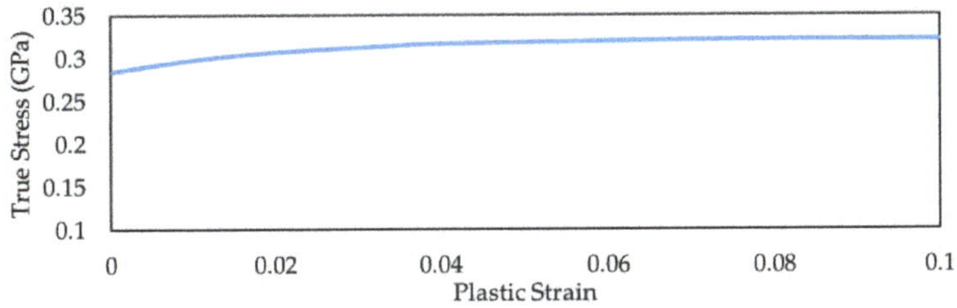

Figure 16. Polymer Lithium-Ion Battery (PLIB) aluminium casing hardening curve.

Table 13. Floor, upper, and lower plates dimensions.

<table>
<tr><th>Floor Modelling</th><th colspan="2">Value</th><th colspan="2">Unit</th></tr>
<tr><td>Dimension ($\hat{x} \times \hat{z}$)</td><td colspan="2">200 × 200</td><td colspan="2">mm</td></tr>
<tr><td>Thickness ($\hat{y}$)</td><td colspan="2">2</td><td colspan="2">mm</td></tr>
<tr><th>Battery Modelling</th><th colspan="2">Value</th><th colspan="2">Unit</th></tr>
<tr><td>Dimension ($\hat{x} \times \hat{z}$)</td><td colspan="2">150 × 92</td><td colspan="2">mm</td></tr>
<tr><td>Thickness ($\hat{y}$)</td><td colspan="2">30</td><td colspan="2">mm</td></tr>
<tr><td>Element model</td><td colspan="2">Fully Integrated Solid</td><td colspan="2">-</td></tr>
<tr><td>Density (ρ)</td><td colspan="2">1.76×10^{-6}</td><td colspan="2">kg/mm^3</td></tr>
<tr><td>Young's Modulus (E)</td><td colspan="2">0.904</td><td colspan="2">GPa</td></tr>
<tr><td>Tensile Cut-off (TSC)</td><td colspan="2">108</td><td colspan="2">MPa</td></tr>
<tr><td>DAMP Factor</td><td colspan="2">0.5</td><td colspan="2">-</td></tr>
<tr><th rowspan="2">Plate Modelling</th><th colspan="2">Upper Plate</th><th colspan="2">Lower Plate</th></tr>
<tr><th>Value</th><th>Unit</th><th>Value</th><th>Unit</th></tr>
<tr><td>Dimension ($\hat{x} \times \hat{z}$)</td><td>208 × 208</td><td>mm</td><td>208 × 208</td><td>mm</td></tr>
<tr><td>Thickness ($\hat{y}$)</td><td>0.4</td><td>mm</td><td>0.25</td><td>mm</td></tr>
</table>

This optimisation has four different cell structure configurations, as shown in Figure 17. The optimised model is used for all cell configurations. Each configuration's element size equals the thickness of a single cell divided by 3. The dimensions of each configuration are detailed in Table 14.

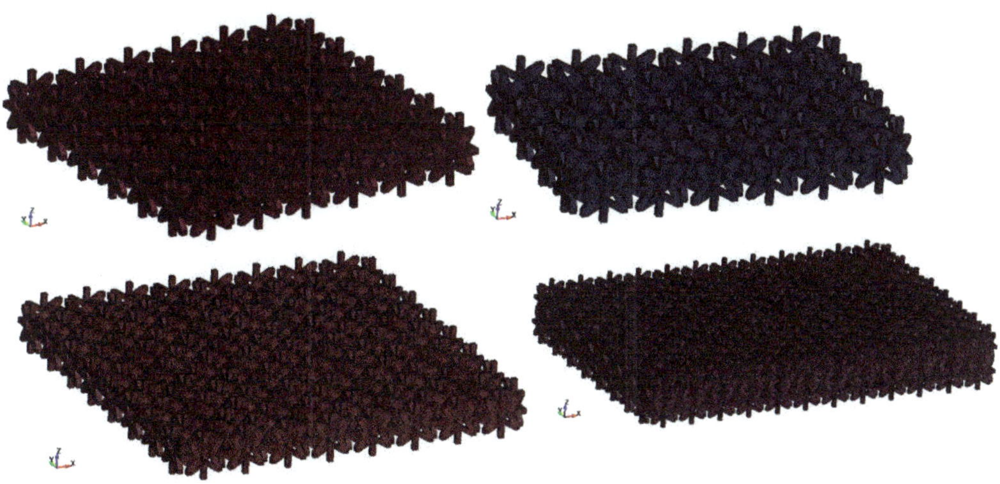

Figure 17. Cell configuration 1 (**upper left**), cell configuration 2 (**upper right**), cell configuration 3 (**lower left**), cell configuration 4 (**lower right**).

Table 14. Cell configurations for battery protection system.

Configuration	Cell Resize	Cell Arrangement	Total Dimension	Structural Volume
1	100%	$6 \times 6 \times 1$	$165.4 \times 165.4 \times 30$ mm	140,271 mm^3
2	100%	$6 \times 4 \times 1$	$165.4 \times 111.2 \times 30$ mm	95,695.6 mm^3
3	80%	$8 \times 8 \times 1$	$175.5 \times 175.5 \times 24$ mm	124,980 mm^3
4	50%	$12 \times 8 \times 2$	$163.9 \times 109.7 \times 30$ mm	89,017.2 mm^3

The impactor simulates an outside item striking the battery system upon ground impact. The impactor was modelled as a flat-tipped cone. It weighs 0.77 kg and is designed to move in the direction of Z+ at an initial speed of 42 m/s. The total energy of the system during simulation is equal to the initial kinetic energy of the impactor, which is 678.74 J.

4.2. Results and Discussion

The numerical simulation's result was applied over a period of 2 ms. Table 15 shows the impact simulation result.

Table 15. Visualisation of the battery thickness deformation for four configurations during impact. The battery is the most deformed for simulation using configuration 3 and least deformed using configuration 1.

Configuration 1	Configuration 2	Configuration 3	Configuration 4	Time
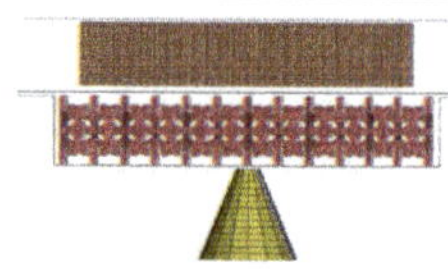	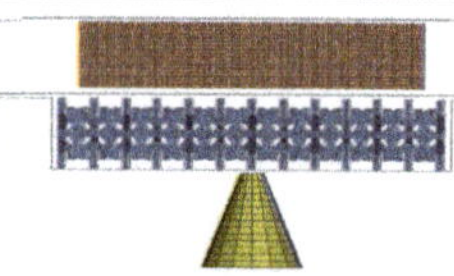	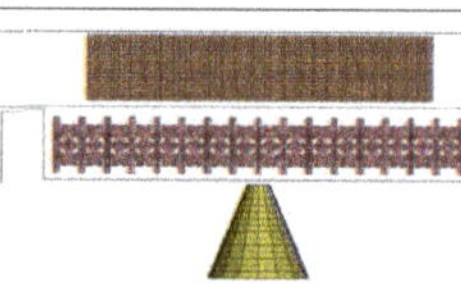	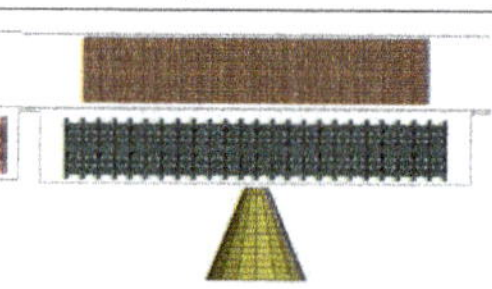	0 ms
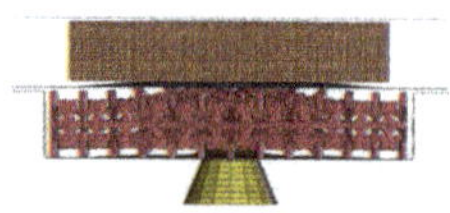	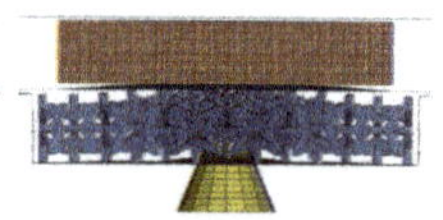		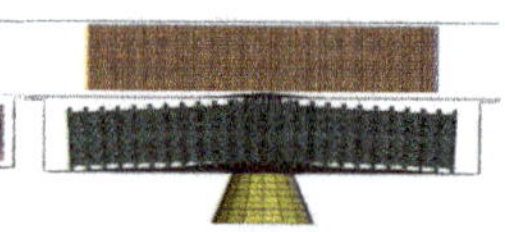	0.5 ms
				1 ms
				1.5 ms
				2 ms

Calculations are made to determine whether the cell structure can minimize battery deformation by measuring the displacement of nodes along the battery. The battery planar, predicted to experience the greatest deformation, is evaluated at the centre. To view the deformation as a time function, the local thickness—the distance between the highest and lowermost nodes—is determined at that point during the whole simulation time interval using Equation (1).

$$Max\,deformation = (z_{upper} - z_{lower}) - 30\ \text{mm}, \tag{1}$$

Firstly, to decide the best configuration, the battery must be shielded from impact loads to function properly throughout the simulation. Secondly, the ideal battery protector should be as compact and lightweight as possible because then a smaller amount of energy will be required to operate the vehicle.

Figure 18 shows the change in maximum deformation for all cell configurations. Only the third configuration fails to protect the battery as it underwent more deformation than is necessary for battery failure. The first, second, and fourth configurations thereby satisfy the first condition. The structure's dimensions and mass are compared to pick between the other three configurations.

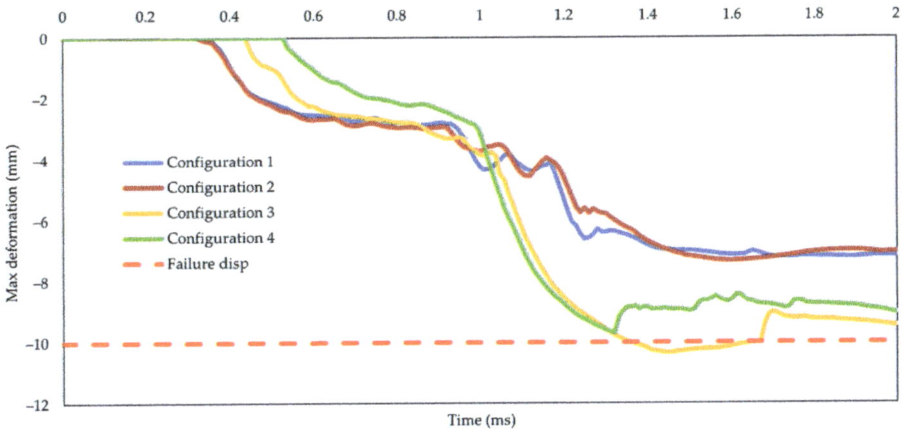

Figure 18. Change in maximum deformation throughout simulation for all cell configurations.

Table 16 shows that the second configuration is 7.5% heavier than the fourth configuration but 46.6% lighter than the first. However, compared to the first and fourth configurations (46.6% and 15.7%, respectively), the cell structure of the second configuration yields the highest SEA value. Based on this number, the second configuration is the most effective for protecting the battery.

Table 16. Simulation results comparison for all cell configurations.

Configuration	Volume (mm^3)	Mass (kg)	EA (J)	SEA (J/kg)	Maximum Deformation (mm)
1	140,271	0.379	225.4	595.2	−7.21
2	95,695.6	0.258	225.3	872.2	−7.33
3	124,980	0.337	126.3	374.4	FAIL
4	89,017.2	0.240	181.1	753.6	−9.69

Smaller single-cell designs tend to experience more deformation, which suggests that the amount of deformation experienced by the battery is related to the single-cell structure's size. Due to the larger single-cell size of the configuration structure, it was predicted that thicker materials possess greater flexural stiffness and are more capable of resisting three-point bending, following Campbell [9]. By comparing the third and fourth configurations, it is possible to see that when modelling many stacks for smaller single-cell structures, the SEA increases. The energy values for the battery system and the second configuration cell structure are also plotted to test the structure's capacity to absorb impact energy.

As shown in Figure 19, the system's total energy (about 679 J) remains constant throughout the simulation. The simulation result in Table 17 shows some damage to the cell structure starting at 1 ms, indicating that the material's failure criterion has been met and some of the cell's elements have begun to erode. The corroded elements are also causing a slight decrease in the energy of the cell structure (Figure 20). For the simulation result to be accurate, the system's sliding energy, the sum of friction energy from all contacts, must be positive. Due to the initial velocity of the impactor, kinetic energy is the only energy in the system. It gradually decreases as internal energy increases, corresponding to the actual physical situation where the impactor is slowed down when interacting with the battery protector.

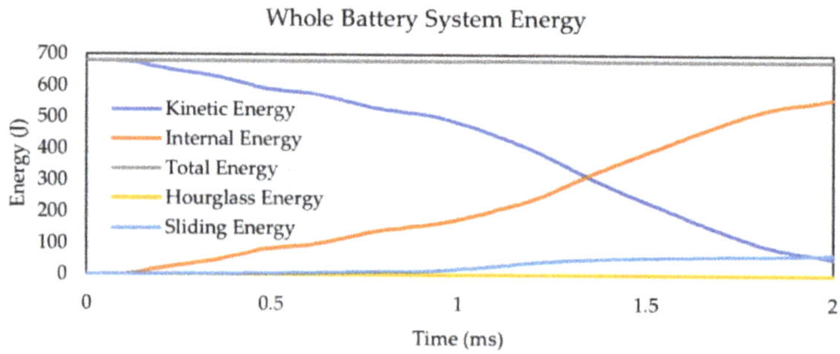

Figure 19. Energy vs. time curve for whole battery system.

Table 17. Numerical simulation results for the optimised cell configuration.

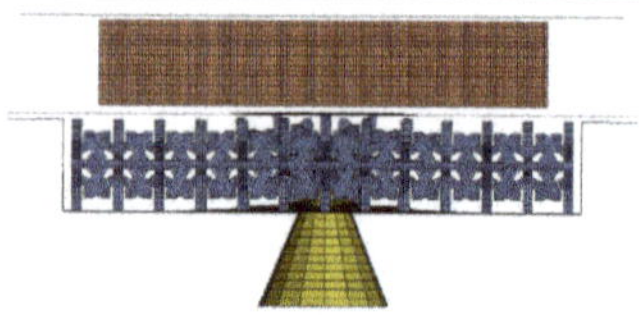	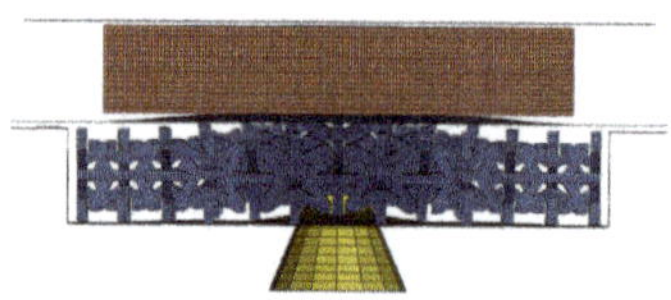	
0.25 ms	0.5 ms	0.75 ms

1 ms	1.25 ms	1.5 ms
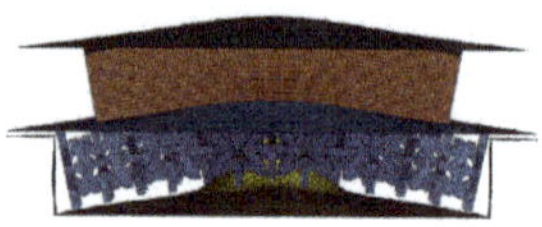		
1.75 ms		2 ms

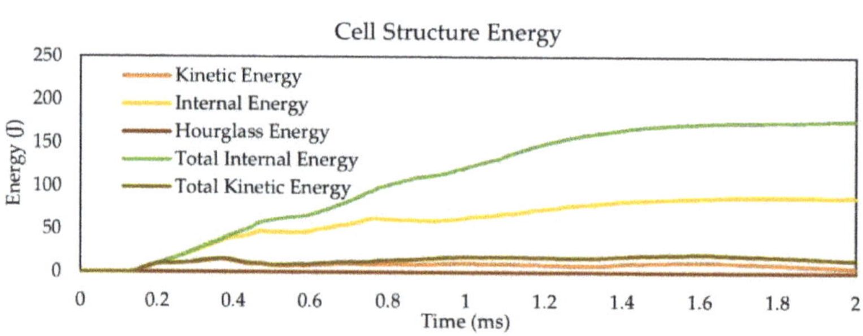

Figure 20. Energy vs. time curve for metastructure cell structure.

A battery failure can be avoided by the optimised structure using configurations 1, 2, and 4, given that no point has been compressed more than 10.4 mm. Considering the dimension, mass, and specific energy consumption value, the second configuration is the best to protect the prismatic battery.

Nevertheless, the cell structure, which is supposed to become denser as compression occurs, could not form its auxetic characteristic due to the crushing trend seen in Figure 21. Instead, the impulsive force caused the cell structure to lift, which may be caused by the concentrated (not uniformly dispersed) impact mechanism at one point.

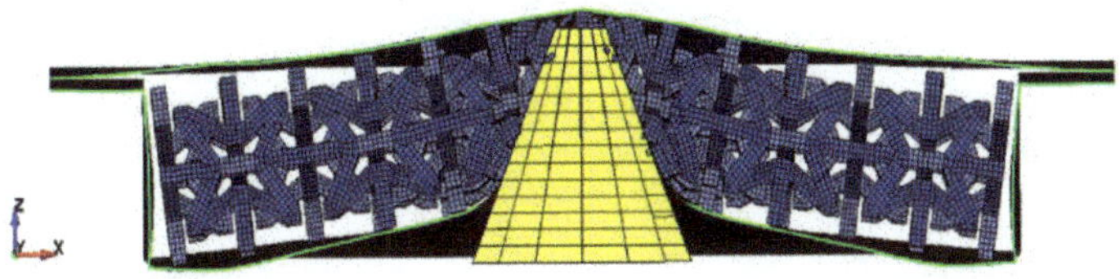

Figure 21. Numerical simulation result of the most optimised battery protector system configuration using concentrated impactor.

On the other hand, the design optimisation for data training utilised a rigid plane as the impactor to yield an even distribution. The cell structure can be scaled up to the point where the side of the cell is wide enough to bend when in contact with enclosing plates to avoid this, resulting in the auxetic feature shown in Figure 22.

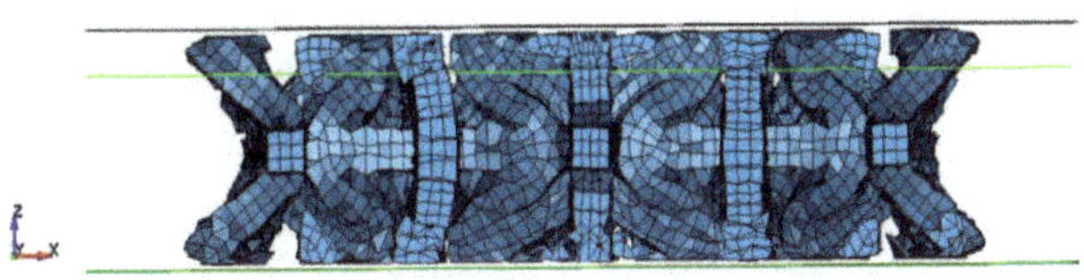

Figure 22. Numerical simulation result of the optimised cell structure for quasi-static loading using rigid plane impactor from data training process.

5. Conclusions

Preliminary modelling and validation of simulation were used to determine the valid modelling parameters and baseline performance using the optimised beam-type bi-stable metastructure from Kirana's research [19]. The simulation has shown that the baseline structure can absorb a crash energy of 58.7 J. The SEA value of the baseline model is 996.9 J/kg, given its mass of 0.0589 kg.

ANN and NSGA-II were used to predict the optimised metastructure configuration. It is found that the star-shaped auxetic cell can absorb the most crushing energy while having the least amount of mass, which results in the highest SEA. When subjected to a uniformly dispersed load, the cell structure has auxetic characteristics due to the crushing trend where the centre of the cell laterally shrinks in the transversal direction, as shown in Figure 22.

The optimal cell (Figure 12) has the following dimensions: 10.2 mm inner spacing, 2.9 mm thickness, and consists of 1 stack. With 0.049 kg of mass and 2772.38 J of absorbed energy, the optimal cell has a SEA value of 56,596.28 J/kg or 5577% higher than the baseline model, according to the numerical simulation results.

Furthermore, the best cell configuration is the second arrangement, based on the whole battery system simulation results and comparison analysis, (Figure 17), in which the optimised cell is arranged in $6 \times 4 \times 1$ cells for a total dimension of $165.4 \times 111.2 \times 30$ mm. This dimension formed a 95,695.6 mm^3 volume with 0.258 kg of mass. With this arrangement, the battery's maximum deformation is 7.33 mm, significantly less than the deformation required for prismatic battery failure (10.423 mm).

Future research is suggested to improve this work by extending the data training process, investigating different machine learning methods, and also validating other shape-type options. The data training process can be developed by considering new geometrical shapes and additional materials such as composites, simulating additional impactor conditions, and other loading scenarios with manufacturing aspects in the simulation. Another optimisation method to be explored can conclude single-objective genetic algorithms that are Pareto-dominance-based.

Author Contributions: Conceptualization, methodology, validation, analysis review, resources, writing—review and editing, supervision, project administration, S.P.S.; Numerical modelling, numerical analysis and validation, investigation, data curation, writing—original draft preparation, writing—review and editing, visualization, I.C.F.; Methodology, validation, analysis review, resources, writing—review and editing, supervision, project administration, D.W. and Methodology, validation, analysis review, resources, writing—review and editing, supervision A.N.P. All authors have read and agreed to the published version of the manuscript.

Funding: The authors received no specific funding from any organisation for the submitted work. No funding was given to assist with the preparation of this manuscript and conducting this study.

Data Availability Statement: Due to confidentiality agreements, data created and analysed during the research is available from the corresponding author on request.

Acknowledgments: The authors would like to acknowledge the Teladan scholarship program from Tanoto Foundation (Jakarta, Indonesia) for providing support in fulfilling the degree of Sarjana Teknik from Institut Teknologi Bandung.

Conflicts of Interest: The authors declare no conflict of interest.

References

1. Fleck, A. EV Market Revenue Set To Hit $384 Billion in 2022. Available online: https://www.statista.com/chart/28211/electric-vehicles-revenue-projections/ (accessed on 14 July 2023).
2. The Global Electric Vehicle Market Overview in 2023: Statistics & Forecasts. Available online: https://www.virta.global/en/global-electric-vehicle-market#one (accessed on 14 July 2023).
3. Batteries for Electric Vehicles. Available online: https://afdc.energy.gov/vehicles/electric_batteries.html (accessed on 14 July 2023).
4. Preventing Fire and/or Explosion Injury from Small and Wearable Lithium Battery Powered Devices. Available online: https://www.osha.gov/sites/default/files/publications/shib011819.pdf (accessed on 14 July 2023).
5. Sahraei, E.; Meier, J.; Wierbicki, T. Characterizing and Modeling Mechanical Properties and Onset of Short Circuit for Three Types of Lithium-Ion Pouch Cells. *J. Power Sources* **2014**, *247*, 503–516. [CrossRef]
6. Chen, Y.; Kang, Y.; Zhao, Y.; Wang, L.; Liu, J.; Li, Y.; Liang, Z.; He, X.; Li, X.; Tavajohi, N.; et al. A Review of Lithium-Ion Battery Safety Concerns: The Issues, Strategies, and Testing Standards. *J. Energy Chem.* **2021**, *59*, 83–99. [CrossRef]
7. Xia, Y.; Wierzbicki, T.; Sahraei, E.; Zhang, X. Damage of Cells and Battery Packs Due to Ground Impact. *J. Power Sources* **2014**, *267*, 78–97. [CrossRef]
8. Yazdani Sarvestani, H.; Akbarzadeh, A.H.; Mirbolghasemi, A.; Hermenean, K. 3D Printed Meta-Sandwich Structures: Failure Mechanism, Energy Absorption and Multi-Hit Capability. *Mater. Des.* **2018**, *160*, 179–193. [CrossRef]
9. Campbell, F.C., Jr. *Manufacturing Technology for Aerospace Structural Materials*; Elsevier: Amsterdam, The Netherlands, 2006; ISBN 978-1-85617-495-4.
10. Bertoldi, K.; Vitelli, V.; Christensen, J.; Van Hecke, M. Flexible Mechanical Metamaterials. *Nat. Rev. Mater.* **2017**, *2*, 17066. [CrossRef]
11. Che, K.; Yuan, C.; Wu, J.; Jerry Qi, H.; Meaud, J. Three-Dimensional-Printed Multistable Mechanical Metamaterials with a Deterministic Deformation Sequence. *J. Appl. Mech.* **2017**, *84*, 011004. [CrossRef]
12. Liu, J.; Chang, J.; Yu, J.; Zhang, W.; Huang, S. Machine Learning-Based Optimization Design of Bistable Curved Shell Structures with Variable Thickness. *Structures* **2023**, *52*, 175–186. [CrossRef]
13. Gibson, L.J.; Ashby, M.F. Cellular Solids: Structure and Properties. Available online: https://www.cambridge.org/core/books/cellular-solids/BC25789552BAA8E3CAD5E1D105612AB5 (accessed on 14 July 2023).
14. Zhang, J.; Lu, G.; You, Z. Large Deformation and Energy Absorption of Additively Manufactured Auxetic Materials and Structures: A Review. *Compos. Part B Eng.* **2020**, *201*, 108340. [CrossRef]
15. Wang, L.; Sun, J.; Ding, T.; Liang, Y.; Ho, J.C.M.; Lai, M.H. Manufacture and Behaviour of Innovative 3D Printed Auxetic Composite Panels Subjected to Low-Velocity Impact Load. *Structures* **2022**, *38*, 910–933. [CrossRef]

16. Ribeiro Filho, S.L.M.; Tonatto, M.L.P.; Hallak Panzera, T.; Remillat, C.D.L.; Scarpa, F. Multi-Objective Optimisation of Aluminium Skins and Recycled/Perforated PET Foams Sandwich Panels Subjected to Impact Loads. *Structures* **2022**, *43*, 1750–1765. [CrossRef]
17. Brendon Francisco, M.; Luiz Junho Pereira, J.; Augusto Vilas Boas Vasconcelos, G.; Simões da Cunha Jr, S.; Ferreira Gomes, G. Multiobjective Design Optimization of Double Arrowhead Auxetic Model Using Lichtenberg Algorithm Based on Metamodelling. *Structures* **2022**, *45*, 1199–1211. [CrossRef]
18. Nasrullah, A.I.H.; Santosa, S.P.; Dirgantara, T. Design and Optimization of Crashworthy Components Based on Lattice Structure Configuration. *Structures* **2020**, *26*, 969–981. [CrossRef]
19. Kirana, S. *Studi Parametrik Konfigurasi Metastruktur Logam Berbasis Open Cell Dengan Pembebanan Kompresi Untuk Aplikasi Kendaraan Tempur*; Institut Teknologi Bandung: Bandung, Indonesia, 2021.
20. Biharta, M.A.S.; Santosa, S.P.; Widagdo, D.; Gunawan, L. Design and Optimization of Lightweight Lithium-Ion Battery Protector with 3D Auxetic Meta Structures. *World Electr. Veh. J.* **2022**, *13*, 118. [CrossRef]
21. Carakapurwa, F.E.; Santosa, S.P. Design Optimization of Auxetic Structure for Crashworthy Pouch Battery Protection Using Machine Learning Method. *Energies* **2022**, *15*, 8404. [CrossRef]
22. Reynolds, C.M.A. Relationship of Mechanical Deformations and Electrochemical Properties of Lithium Ion Batteries—An Experimental Study. Available online: https://dspace.mit.edu/bitstream/handle/1721.1/144746/reynolds-cmar14-NavE-MechE-2022-thesis.pdf?sequence=1&isAllowed=y (accessed on 1 March 2023).
23. Zhang, W.; Zhao, S.; Sun, R.; Scarpa, F.; Wang, J. In-Plane Mechanical Behavior of a New Star-Re-Entrant Hierarchical Metamaterial. *Polymers* **2019**, *11*, 1132. [CrossRef]
24. Yang, H.; Wang, B.; Ma, L. Mechanical Properties of 3D Double-U Auxetic Structures. *Int. J. Solids Struct.* **2019**, *180–181*, 13–29. [CrossRef]
25. Zhang, Y.; Tichem, M.; Keulen, F.V. Rotational Snap-through Behavior of Multi-Stable Beam-Type Metastructures. *Int. J. Mech. Sci.* **2021**, *193*, 106172. [CrossRef]
26. AISI Type 304 Stainless Steel. Available online: https://www.matweb.com/search/datasheet.aspx?matguid=bd20a4281ae3430d97cfbebf6904ec50&ckck=1 (accessed on 14 July 2023).
27. ASM Material Data Sheet. Available online: https://asm.matweb.com/search/SpecificMaterial.asp?bassnum=ma6061t6 (accessed on 14 July 2023).
28. García, R.; López, V.-H.; Natividad, C.; Ambriz, R.-R.; Salazar, M. Fusion Welding with Indirect Electric Arc. In *Arc Welding*; IntechOpen: London, UK, 2011; pp. 21–44, ISBN 978-953-307-642-3.
29. Abotula, S.; Chalivendra, V. An Experimental and Numerical Investigation of the Static and Dynamic Constitutive Behaviour of Aluminium Alloys. *J. Strain Anal. Eng. Des.* **2010**, *1*, 1–11. [CrossRef]
30. Joseph, V.R. Optimal Ratio for Data Splitting. *Stat. Anal. Data Min. ASA Data Sci. J.* **2022**, *15*, 531–538. [CrossRef]
31. Sinharay, S. Discrete Probability Distributions. In *International Encyclopedia of Education*, 3rd ed.; Peterson, P., Baker, E., McGaw, B., Eds.; Elsevier: Oxford, UK, 2010; pp. 132–134, ISBN 978-0-08-044894-7.
32. McKay, M.D.; Beckman, R.J.; Conover, W.J. A Comparison of Three Methods for Selecting Values of Input Variables in the Analysis of Output from a Computer Code. *Technometrics* **1979**, *21*, 239–245. [CrossRef]
33. Mckay, M.D.; Beckman, R.J.; Conover, W.J. A Comparison of Three Methods for Selecting Values of Input Variables in the Analysis of Output from a Computer Code. *Technometrics* **2000**, *42*, 55–61. [CrossRef]
34. Gholamy, A.; Kreinovich, V.; Kosheleva, O. Why 70/30 or 80/20 Relation between Training and Testing Sets: A Pedagogical Explanation. In *Departmental Technical Reports (CS)*; 2018; pp. 1–6. Available online: https://scholarworks.utep.edu/cs_techrep/1209 (accessed on 2 January 2023).
35. Tan, H.; He, Z.; Li, E.; Cheng, A.; Chen, T.; Tan, X.; Li, Q.; Xu, B. Crashworthiness Design and Multi-Objective Optimization of a Novel Auxetic Hierarchical Honeycomb Crash Box. *Struct. Multidiscip. Optim.* **2021**, *64*, 2009–2024. [CrossRef]
36. Irawan, D.; Santosa, S.P.; Jusuf, A.; Sambegoro, P.L. Sandwich Panel Composite Based Light-Weight Structure Design for Reserved Energy Storage System (RESS) Protection. In Proceedings of the 2019 6th International Conference on Electric Vehicular Technology, Bali, Indonesia, 18–21 November 2019.
37. Qiao, H.; Wei, Q. 10-Functional Nanofibers in Lithium-Ion Batteries. In *Functional Nanofibers and Their Applications*; Wei, Q., Ed.; Woodhead Publishing Series in Textiles; Woodhead Publishing: Sawston, UK, 2012; pp. 197–208, ISBN 978-0-85709-069-0.

batteries

MDPI

Article

One-Time Prediction of Battery Capacity Fade Curve under Multiple Fast Charging Strategies

Xiaoming Han *, Zhentao Dai *, Mifeng Ren, Jing Cui and Yunfeng Shi

College of Electrical and Power Engineering, Taiyuan University of Technology, Taiyuan 030024, China; renmifeng@tyut.edu.cn (M.R.); cuijing0408@link.tyut.edu.cn (J.C.); shiyunfeng0473@link.tyut.edu.cn (Y.S.)
* Correspondence: hanxiaoming@tyut.edu.cn (X.H.); daizhentao0399@link.tyut.edu.cn (Z.D.)

Abstract: Using different fast charging strategies for lithium-ion batteries can affect the degradation rate of the batteries. In this case, predicting the capacity fade curve can facilitate the application of new batteries. Considering the impact of fast charging strategies on battery aging, a battery capacity degradation trajectory prediction method based on the TM-Seq2Seq (Trend Matching—Sequence-to-Sequence) model is proposed. This method uses data from the first 100 cycles to predict the future capacity fade curve and EOL (end of life) in one-time. First, features are extracted from the discharge voltage-capacity curve. Secondly, a sequence-to-sequence model based on CNN, SE-net, and GRU is designed. Finally, a trend matching loss function is designed based on the common characteristics of capacity fade curves to constrain the encoding features of the sequence-to-sequence model, facilitating the learning of the underlying relationship between inputs and outputs. TM-Seq2Seq model is verified on a public dataset with 132 battery cells and multiple fast charging strategies. The experimental results indicate that, compared to other popular models, the TM-Seq2Seq model has lower prediction errors.

Keywords: lithium-ion battery; multiple fast charging strategies; capacity fade curve; one-time prediction; trend matching

1. Introduction

With the improvement of technology and the decrease of cost, lithium-ion batteries have gradually been promoted. However, their performance will accelerate to fade with the increase of usage times. This fade is the consequence of multiple coupled degradation mechanisms, including electrode–electrolyte interface side reactions, lithium deposition, electrode material structure damage and electrolyte decomposition [1]. Battery capacity is considered the leading indicator for evaluating battery performance. The battery capacity fade curve refers to the trend curve as the battery capacity gradually decreases with the increase of battery usage time. Generally, the battery capacity fade curve shows an exponential fade trend, that is, the longer the battery usage time, the faster the loss of battery capacity. Predicting the battery capacity fade curve can help people understand the battery life and performance, so that they can better manage, use, and maintain the battery. Therefore, it is necessary to develop models that can predict the trend of battery capacity fade using limited data.

Various models have been proposed by researchers to predict the fade trend of battery capacity. These models fall into two main categories: electrochemical models and data-driven models based on battery measurements such as capacity, voltage [2]. Electrochemical models [3] integrate pseudo-two-dimensional (P2D) models [4] or single-particle models [5] with battery aging mechanism models to predict battery capacity fade trends. This approach utilizes partial differential equations [6] to describe physical phenomena like mass transfer, charge transfer, thermodynamics, and chemical kinetics [7,8]. Although electrochemical models can provide highly accurate prediction results in some cases, there are many uncertainties in the electrochemical process under different usage conditions, such as temperature, pressure, depth of charge–discharge, etc. These factors can affect the internal structure and chemical reactions of the battery that result in modeling failure.

Data-driven models have been widely applied. Some researchers used historical capacity data as input and fitted the capacity fade situation by using mathematical models such as exponential [9] and double exponential [10]. Other researchers consider more diverse inputs and employ deep learning models to capture capacity fade trends [11–14]. The aforementioned studies typically required sufficient charge–discharge data to accurately predict future capacities, which has limited the application of battery capacity prediction techniques in many scenarios.

Li et al. [15] implemented the earliest prediction of the entire capacity fade trend using a capacity-based method. This method relies solely on the capacity series data as input. They used data up to the first 100 cycles as input and used a sequence-to-sequence model based on a cyclic neural network. While this method predicts the future capacity fade curve in one-time, it fails to consider the impact of various charging conditions. Furthermore, certain high-performance batteries exhibit no capacity deterioration within the first 100 cycles [16]. Therefore, relying solely on capacity as a feature is insufficient for accurately estimating future capacity fade trajectories. It is necessary to extract more comprehensive aging features from data such as voltage, charge quantity, and other data.

Currently, considering the impact of fast charging strategies on batteries and the absence of capacity fade in the initial charge–discharge cycles, some methods have been successfully applied to predict end of life (EOL) [16–18], remaining useful life (RUL) [19], and rollover cycle (knee point) prediction [20]. These methods extract richer aging features from voltage, charge quantity, and other data, enabling the prediction of EOL, RUL and rollover cycle, and promoting the development of capacity degradation trajectory prediction methods. Building upon the various aging features validated by the aforementioned methods, researchers have demonstrated methods for predicting the capacity fade curve by using data from early charge–discharge cycles. Liu et al. [21] proposed an elastic net-based method to estimate parameters in an empirical power function that describes the capacity fade curve over more than 100 cycles. Similarly, Saurabh et al. [22] trained a 2D convolutional model to predict unknown parameters in the capacity fade curve parameter equation. Both of these methods can predict the capacity degradation trajectory. However, the parameter equation cannot fit all the original data without bias. This creates inherent errors in the constructed predicted labels, which can result in a low level of accuracy in the final outcomes.

Strange and dos Reis [23] used two different nine-layer 1D-CNN neural networks to predict the coordinates of multiple sampling points on the capacity degradation curve. Similarly, Herring et al. [24] used the first 100 cycles of data to build a multitask linear model that predicted the capacity fade curve. Training machine learning models using small amounts of battery data still presents a challenge, despite the absence of errors in label construction. Overfitting commonly plagues these approaches due to three primary reasons stemming from the nature of the data itself. Firstly, there are significant differences between samples, leading to complexities in modeling. Secondly, each sample exhibits high dimensions, with input features comprising multiple time series and the output label consisting of a single time series. Moreover, the resource-intensive nature of full-cycle experiments for batteries results in limited available training data. Consequently, fitting the training data necessitates larger models. However, without considering the distribution characteristics of the data, these models often lack adequate generalization ability. This predicament raises the unresolved issue of how to develop models with strong generalization capabilities in the face of minimal battery data.

In order to address the aforementioned issues, a battery capacity degradation trajectory prediction method based on the TM-Seq2Seq (Trend Matching—Sequence-to-Sequence) model is proposed. This method predicts the future capacity fade curve in one-time without iteration. Firstly, features are extracted during the first 100 charge–discharge cycles. Secondly, the capacity fade curve is sampled at equal capacity intervals, and the sampled sequence is set as labels. Next, features are extracted using dilated convolution and SE-net, and the features are enhanced using a GRU decoder. Finally, a GRU decoder is used to predict the labels. Additionally, a trend feature matching loss function is designed based on the data characteristics of capacity fade curves. This loss function constrains the output from the intermediate layers of the model, ensuring that the intermediate features follow a certain pattern similar to the capacity fade curve. By controlling the learning space in this way, the model is able to enhance its robustness and improve its learning capabilities. Finally, a parameter equation for the capacity degradation curve is designed to fit the predicted sample points and obtain the complete capacity degradation curve. The proposed method is verified on a public dataset with 132 battery cells and multiple fast charging strategies.

2. The Battery Data for Multiple Fast Charging Strategies

Different fast charging strategies can impact the degradation rate of batteries. To reflect this scenario, this study utilizes the graphite/LiFePO4 cylindrical batteries cycle dataset from Severson et al. [16] and Attia et al. [25]. The experimental dataset consists of 132 batteries of the same type with different fast charging strategies. Figure 1 shows six fast charging strategies' current-capacity curves and voltage-capacity curves. Firstly, these fast charging strategies involve constant current charging at different rates in different capacity ranges. Secondly, each battery follows the same fast charging strategy throughout different charge–discharge cycles. Lastly, all batteries have the same discharge conditions. Figure 2a displays the capacity fade curves of the 132 batteries. The horizontal axis represents the cycle number, the vertical axis represents the battery capacity, and each curve represents the capacity fade curve of a single battery. The color of the curve reflects the lifespan of the battery. The distribution of the total test cycles of 132 batteries is illustrated in Figure 2b. Under the influence of different charging strategies, the capacity degradation rate and trend of the batteries exhibit uncertainty, posing challenges to the prediction task of capacity fade curves.

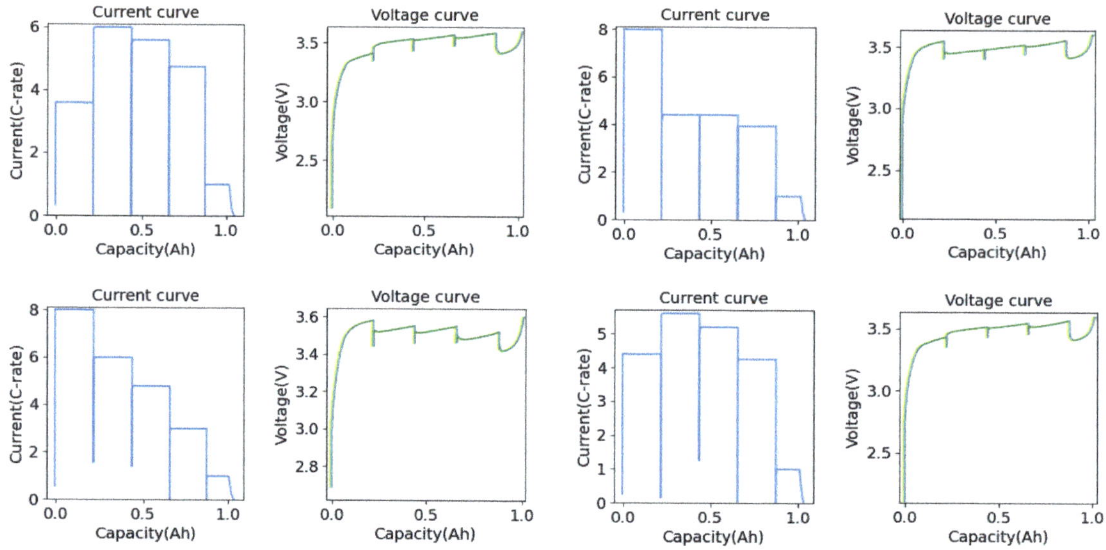

Figure 1. *Cont.*

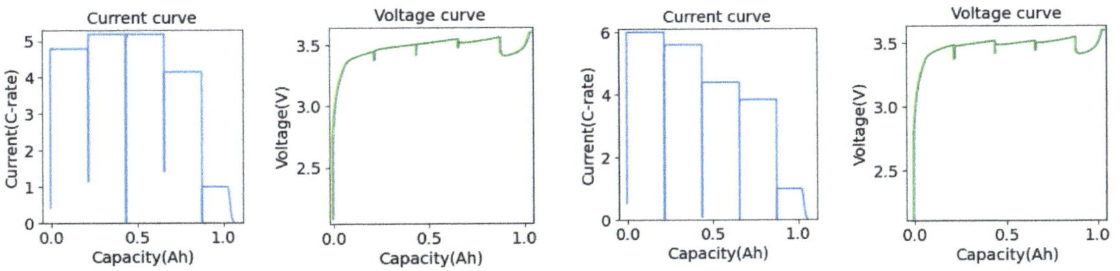

Figure 1. Current and voltage curves for six different fast charging strategies.

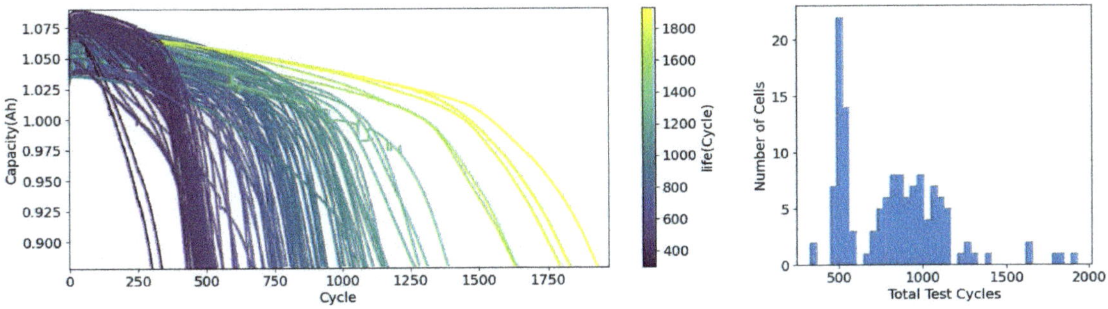

(**a**) (**b**)

Figure 2. (**a**) Capacity fade curves of 132 experimental batteries. (**b**) The distribution of the total test cycles of 132 batteries.

Further information about the dataset can be found at the following web address:

1. https://data.matr.io/1/projects/5c48dd2bc625d700019f3204, accessed on 1 January 2024.
2. https://data.matr.io/1/projects/5d80e633f405260001c0b60a, accessed on 1 January 2024.

3. One-Time Prediction Method for Capacity Fade Curves

In this section, we explore a comprehensive approach to predict the capacity fade curve. This section comprises three subsections, each addressing a specific aspect of the methodology. Section 3.1 provides an overview of the feature extraction method as well as the method of quantifying the capacity fade curve into labels. Section 3.2 delves into the principles of the TM-Seq2Seq model, which serves as the core framework for our prediction methodology. Lastly, Section 3.3 presents the capacity fade curve parameter equations used to fit the labels into smooth and accurate representations of the actual fade curves.

3.1. Feature Structure and Label Structure

Although multiple fast charging strategies greatly affect the rate of battery capacity degradation, most batteries do not exhibit significant capacity fade within the first 100 charge–discharge cycles, as shown in Figure 3a. Relying solely on capacity data cannot reflect the battery's aging rate. Therefore, it is necessary to identify features that reflect the battery's aging rate and trend and map them onto future capacity fade curves. The voltage–discharge capacity curve is an important curve that reflects the discharge voltage strength of the battery at different states of charge. Many studies have extracted battery aging features through Q(V) (discharge capacity–voltage) curves and conducted a series of prediction tasks. As the battery ages, the Q(V) curve of the battery will undergo changes. Figure 3b illustrates the differences between the Q(V) curves of the battery during the first cycle and the 100th cycle. The horizontal axis of the graph represents the variance in capacity between the first cycle and the 100th cycle, while the vertical axis depicts the discharge voltage of the battery.

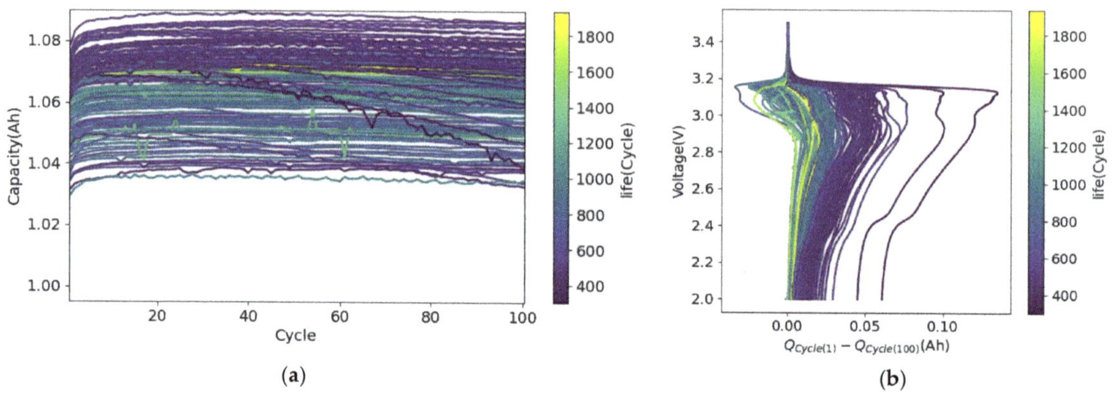

Figure 3. (**a**) Capacity curves within the first 100 charge–discharge cycles. (**b**) The subtraction of the Q(V) curve of the first cycle from the Q(V) curve of the 100th cycle of a battery.

The extracted features need to reflect the changes in the discharge Q(V) curve. Considering this aspect, two sets of temporal features are extracted. The specific feature extraction method is as follows.

Firstly, to ensure accurate feature extraction, it is important to establish a specific voltage range. The high-rate constant current discharge often leads to a reduction in the initial voltage, which can create difficulties in measuring voltages within the 3.5 V to 3.6 V range. Consequently, to address these challenges, we have opted to focus on extracting features within the voltage span of 2.0 V (discharge cut-off voltage) to 3.5 V. The discharge capacity values between 2.0 V and 3.5 V on the Q(V) curve are sampled at equal voltage intervals, generating a vector $\boldsymbol{Q}_i = \{Q_i(V_{\min}), \ldots, Q_i(V_{\max})\}$ composed of the discharge capacity values. In this vector, the i is the cycle number and $\boldsymbol{Q}_i(V_{\min})$ is the discharge capacity at a voltage of $V_{\min}$.

Next, the difference of vector $\boldsymbol{Q}_i$ for each cycle relative to the 100th cycle is calculated using Equation (1). This difference serves to reflect the variations in the Q(V) curve.

$$\triangle \boldsymbol{Q}_{(100-i)} = \boldsymbol{Q}_{100} - \boldsymbol{Q}_i \tag{1}$$

Lastly, we use the mean and variance of the vector $\triangle \boldsymbol{Q}_{(100-i)}$ to construct two features. Specifically, the variance feature is calculated as shown in Equation (2), while the mean feature is computed following Equation (3).

$$\begin{cases} \boldsymbol{Var}_j = \left\{ Var\left(\triangle \boldsymbol{Q}_{(100-1)j}\right), Var\left(\triangle \boldsymbol{Q}_{(100-2)j}\right), \ldots, Var\left(\triangle \boldsymbol{Q}_{(100-100)j}\right) \right\} \\ Var\left(\triangle \boldsymbol{Q}_{(100-i)}\right) = \frac{1}{n} \sum\limits_{i=1}^{n} \left(\triangle \boldsymbol{Q}_{(100-i)}(V_i) - \overline{\triangle \boldsymbol{Q}_{(100-i)}}\right)^2 \end{cases} \tag{2}$$

$$\boldsymbol{Mean}_j = \left\{ \overline{\triangle \boldsymbol{Q}_{(100-1)j}}, \overline{\triangle \boldsymbol{Q}_{(100-2)j}}, \ldots, \overline{\triangle \boldsymbol{Q}_{(100-100)j}} \right\} \tag{3}$$

where $\boldsymbol{Mean}_j$ and $\boldsymbol{Var}_j$ is the features, j represents the j-th battery, which also corresponds to the j-th sample. These features reflect the speed and trend of battery aging within the first 100 cycles.

When the actual capacity of the battery reaches 80% of its rated capacity, the battery life ends. The end-of-life capacity of the battery can be calculated based on the rated capacity of the battery, as shown in Equation (4).

$$C_{EOL} = C_{\text{Nominal}} \times 80\% \tag{4}$$

where C_{Nominal} is the nominal capacity of the battery, which is 1.1 Ah, and C_{EOL} is the end-of-life (EOL) capacity of the battery.

To achieve a reliable prediction of the capacity decay trajectory from the 100th cycle to end of life, we establish the target label that accurately reflects this progression. In order to maintain uniform label length across samples, we choose to evenly select 10 points along the capacity decay curve. This decision is driven by the intention to mitigate the risk of overfitting the model due to high label dimensions, particularly when dealing with limited sample data. Therefore, 10 points are uniformly sampled between C_{100} and C_{EOL}. These points can be obtained using Equation (5).

Next, to determine the coordinates of these 10 capacity values, we simply need to acquire their respective cycle numbers. The cycle numbers corresponding to these 10 capacity values are set as the labels for prediction, as shown in Equation (6).

$$\begin{cases} C_{ij} = C_{100j} - \left(C_{100j} - C_{EOL}\right) \times i \times 10\% \\ C_j = \left\{C_{1j}, C_{2j}, \ldots, C_{10j}\right\} \end{cases} \tag{5}$$

$$\boldsymbol{y}_j = \left\{Cycle_{1j}, Cycle_{2j}, \ldots, Cycle_{10j}\right\} \tag{6}$$

where $Cycle_{ij}$ is the number of cycles to reach the capacity value of C_{ij} (specifically, $C_{10j} = C_{EOL}$, and $Cycle_{10j}$ is the EOL). Vector $\boldsymbol{y}_j$ is the label for prediction. J represents the j-th battery, which also corresponds to the j-th sample.

The initial measured capacity of different batteries deviates from the nominal capacity, resulting in variations in the numerical range of Vector $\boldsymbol{C}_j$ for different batteries. Additionally, the elements in Vector $\boldsymbol{C}_j$ and Vector $\boldsymbol{y}_j$ affect each other mutually. Given the above reasons, vector $\boldsymbol{C}_j$ is used to predict the label $\boldsymbol{y}_j$. In summary, the features to be inputted into the model and the label to be predicted are shown in Equation (7).

$$D = [input|target] = \left[\begin{array}{ccc|c} \boldsymbol{Var}_1, & C_1, & \boldsymbol{Mean}_1 & y_1 \\ \boldsymbol{Var}_2, & C_2, & \boldsymbol{Mean}_2 & y_2 \\ & \cdots & & \cdots \\ \boldsymbol{Var}_n, & C_n, & \boldsymbol{Mean}_n & y_n \end{array}\right] \tag{7}$$

3.2. The Structure of TM-Seq2Seq Model

The original feature sequences used to predict the capacity fade curve represent a high-dimensional input for the sequence-to-sequence regression task. However, directly feeding these sequences into a Recurrent Neural Network (RNN) could hinder the GRU's ability to retain information over extended periods. Furthermore, the time series features being inputted into the model hold diverse physical interpretations, necessitating a careful evaluation of their relative importance for improving prediction accuracy. Therefore, to address these challenges, the TM-Seq2Seq model structure was specifically crafted with a focus on the sequence-to-sequence architecture [26] as shown in Figure 4. Firstly, the original features are learned and dimensionality-reduced through strided convolutions. Following this, an SE-net is added after the convolutional neural network to assign weights to features derived from various convolutional channels. Then, an encoder composed of GRU is used to encode the input sequence. Finally, a decoder composed of GRU is used to predict the labels $\boldsymbol{y}_j$. In addition, the elements in Vector $\boldsymbol{C}_j$ and Vector $\boldsymbol{y}_j$ affect each other mutually. Given the above reasons, vector $\boldsymbol{C}_j$ is used as input for the decoder. Furthermore, considering that the capacity fade curve has characteristics approximating exponential fade, the trend features of different sample encoding vectors are constrained. The specific roles and principles of SE-net, GRU, and trend matching are introduced below.

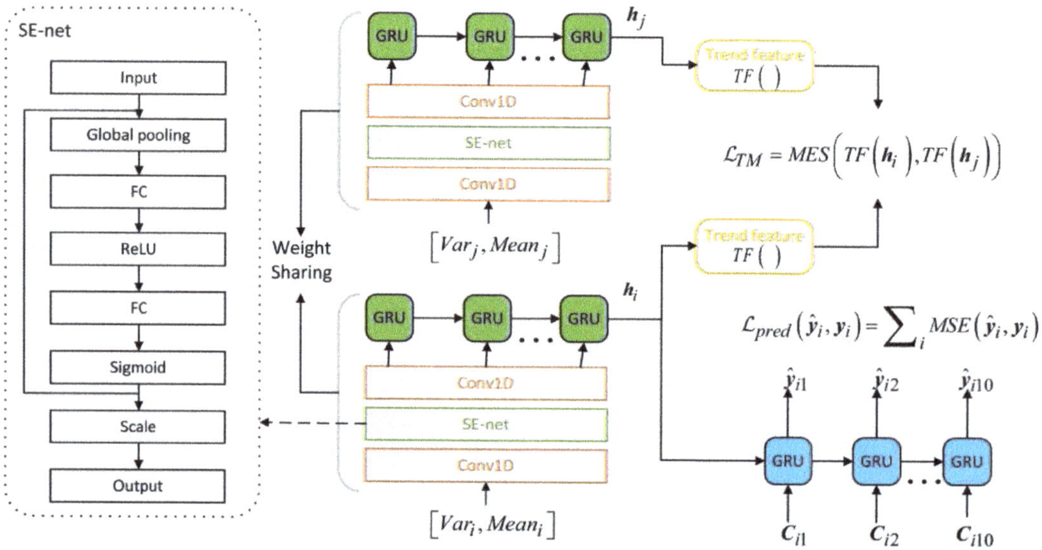

Figure 4. The structure of TM-Seq2Seq model.

3.2.1. SE-net

SE-Net (Squeeze-and-Excitation Network) [27] is an attention mechanism used to enhance the performance of convolutional neural networks (CNN). By introducing channel attention mechanism, SE-Net enables the network to dynamically learn the importance of each channel. First, for the input features, SE-Net performs global average pooling operation to transform the features of each channel into a single value. This value represents the global statistical information of that channel's feature. Second, SE-Net introduces a fully connected two-layer network to learn the weight of each channel. Finally, by multiplying the learned channel attention weights with the original features, the effect of strengthening Important features and weakening useless features is achieved. The formula for the "squeeze" operation is shown in Equation (8).

$$\begin{cases} z_c = F_{sq}(u_c) = \frac{1}{W\times H}\sum_{i=1}^{W}\sum_{j=1}^{H} u_c(i,j) \\ U = (u_1, u_2, \ldots, u_c) \end{cases} \tag{8}$$

where U is a set of input features, u_c is an input sample, and z_c is global statistical information.

The "excitation" operation is shown in Equation (9). The channel dependency is obtained through a simple gating mechanism.

$$s_c = F_{ex}(s_c, W) = \sigma(g(z_c, W)) = \sigma(W_2\delta(z_c W_1)) \tag{9}$$

where W_1 and W_2 are the weight matrix to be optimized. s_c represents the channel weight vector.

The output of the SE-net is shown in Equation (10).

$$\widetilde{x}_c = F_{scale}(u_c, s_c) = s_c \cdot u_c \tag{10}$$

where F_{scale} represents the multiplication operation channel-wise.

3.2.2. GRU

GRU [28] (Gated Recurrent Unit) comprises two gates: the update gate and the reset gate. The reset gate determines the influence of the old hidden state on the current state, while the update gate controls the influence of the new input on the current state. The updated equation is as follows:

$$z_t = \sigma(\boldsymbol{W}^z \cdot \boldsymbol{x}_t + \boldsymbol{U}^z \cdot \boldsymbol{h}_{t-1} + \boldsymbol{b}^z) \tag{11}$$

$$\boldsymbol{r}_t = \sigma(\boldsymbol{W}^r \cdot \boldsymbol{x}_t + \boldsymbol{U}^r \cdot \boldsymbol{h}_{t-1} + \boldsymbol{b}^r) \tag{12}$$

$$\boldsymbol{h}'_t = \tanh\left(\boldsymbol{W}^{h'} \cdot \boldsymbol{x}_t + \boldsymbol{U}^{h'} \cdot (\boldsymbol{r}_t \odot \boldsymbol{h}_{t-1}) + \boldsymbol{b}^{h'}\right) \tag{13}$$

$$\boldsymbol{h}_t = (1 - z_t) \odot \boldsymbol{h}_{t-1} + z_t \odot \boldsymbol{h}'_t \tag{14}$$

where $\boldsymbol{x}_t$, $\boldsymbol{h}_t$, and $\boldsymbol{h}'_t$ are input data, output hidden states, and candidate hidden states at time t. The parameters in the weight coefficient matrices, including $\boldsymbol{W}^z$, $\boldsymbol{W}^r$, and $\boldsymbol{W}^{h'}$ for the input data and $\boldsymbol{U}^z$, $\boldsymbol{U}^r$, and $\boldsymbol{U}^h$ for the hidden states, are essential for the training of the network. The biases $\boldsymbol{b}^z$, $\boldsymbol{b}^r$, and $\boldsymbol{b}^h$ are included for each corresponding component. Compared to LSTM, GRU utilizes fewer parameters, which results in faster training and testing speeds. Additionally, the simpler structure helps alleviate the problem of vanishing gradients, making GRU potentially perform better than LSTM when dealing with long-sequence data.

3.2.3. Trend Matching

To effectively fit complex and diverse input–output data, it is crucial to impose appropriate constraints on the model design, given the limited amount of training data available and the significant differences between samples. In designing a method to constrain the model, it is crucial to consider the factors that can impact its learning space. The constraints implemented should not alter the model's structure but rather enhance its generalization ability by limiting the range of parameters it can learn from the data. In many domains, special loss functions are used in generalization techniques to limit the output data from the intermediate layers of the network, thus restricting the learning space of the model [29,30]. The loss function plays a pivotal role in this process as it must be carefully designed to correspond to the data conditions and expected objectives. Given these factors, the method to constrain the model is designed as follows:

Firstly, based on the commonality that the capacity degradation curve roughly follows an exponential fade trend, Equation (15) is designed to quantify the trend feature of a vector.

$$\begin{cases} \boldsymbol{y} = (Cycle_1, Cycle_2, \cdots, Cycle_{10}) \\ TF(\boldsymbol{y}) = \left(\frac{|Cycle_1 - \overline{\boldsymbol{y}}|}{\frac{1}{n}\sum\limits_{i=1}^{n}|Cycle_i - \overline{\boldsymbol{y}}|}, \frac{|Cycle_2 - \overline{\boldsymbol{y}}|}{\frac{1}{n}\sum\limits_{i=1}^{n}|Cycle_i - \overline{\boldsymbol{y}}|}, \cdots, \frac{|Cycle_n - \overline{\boldsymbol{y}}|}{\frac{1}{n}\sum\limits_{i=1}^{n}|Cycle_i - \overline{\boldsymbol{y}}|} \right) \end{cases} \tag{15}$$

where $\boldsymbol{y}$ represents the label and $TF(\boldsymbol{y})$ represents the trend feature of the label.

Secondly, the labels of different batteries were compared and the trend features of these labels were analyzed, as shown in Figure 5. Despite significant variations in the labels of different batteries, they exhibit similar trends. By imposing constraints on different feature encodings, we are able to follow certain patterns with relatively low levels of disorder in the different labels. This approach can help constrain the confusion level associated with various feature encodings, thus making the relationship between feature encodings and labels clearer and more effective. As a result, by enhancing the clarity and effectiveness of this relationship, we can significantly improve the performance and learning capability of the model.

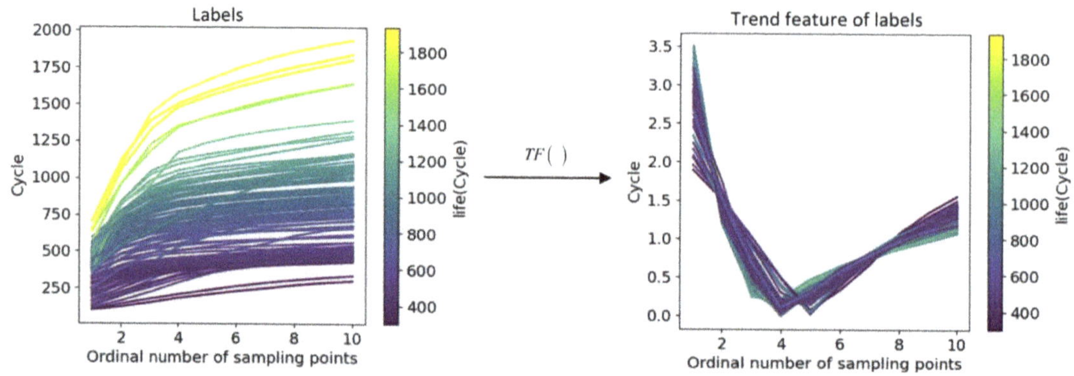

Figure 5. The distribution of the labels and the trend features of these labels.

Finally, Equation (16) is used as the loss function during model training. The purpose of this loss function is to minimize the differences in trend features of feature encodings across different samples. In addition, Equation (19) is also used as a loss function to fit the labels.

$$\mathcal{L}_{TM}(\boldsymbol{h}_i, \boldsymbol{h}_j) = \sum_{i,j} mse(TF(\boldsymbol{h}_i), TF(\boldsymbol{h}_j)) \tag{16}$$

$$\boldsymbol{h}_i = F_{encoder}(\boldsymbol{Mean}_i, \boldsymbol{Var}_i) \tag{17}$$

$$\hat{\boldsymbol{y}}_i = F_{dncoder}(\boldsymbol{h}_i, \boldsymbol{C}_i) \tag{18}$$

$$\mathcal{L}_{pred}(\hat{\boldsymbol{y}}_i, \boldsymbol{y}_i) = \sum_i mse(\hat{\boldsymbol{y}}_i, \boldsymbol{y}_i) \tag{19}$$

where $F_{encoder}$ is the feature extractor consisting of CNN, SE-net, and GRU, $F_{dncoder}$ is the GRU decoder, $\boldsymbol{h}_i$ is the encoding vector of the i-th battery.

3.3. Parameter Equation of Capacity Fade Curve

When parameterizing the capacity fade curve, a reasonable parameter equation should be selected according to the actual situation to ensure accuracy and reliability. The exponential model is a common mathematical model that is used by many studies to fit battery capacity curve. In the dataset of this paper, some batteries experience capacity fade in the first 100 cycles, and full utilization of the known information from these cycles is necessary, such as the initial capacity fade rate. Therefore, we use the sum of the initial capacity fade rate m_0 and the exponential model to represent the change in battery capacity fade rate, where m_0 is the capacity fade rate of the battery in the 100th cycle, and the exponential model can describe the change in battery capacity fade rate. Specifically, the change in capacity fade rate is shown in Equation (20).

$$\frac{dC}{dn} = -m_0 - k \times e^{(n-N_k)\times\delta} \tag{20}$$

where n represents the number of cycles that are experienced by the battery after 100 cycles, m_0 is the capacity fade rate of the battery in the 100th cycle, dC/dn represents the capacity fade rate of the battery in the $100 + n$ th cycle, and k, N_k, and δ are three parameters to be optimized. Integrating Equation (20) yields the relationship between C and n as shown in Equation (21).

$$C = -m_0 \times n - k \times \frac{e^{(n-N_k)\times\delta}}{\delta} + C_1 \tag{21}$$

When $n = 0$, $C = C_{100}$ (C_{100} is the capacity in the 100th cycle). Substituting into Equation (21) can determine the constant C_1:

$$C_{100} = -m_0 \times 0 - k \times \frac{e^{(0-N_k)\times\delta}}{\delta} + C_1 \tag{22}$$

$$C_1 = k \times \frac{e^{-N_k\times\delta}}{\delta} + C_{100} \tag{23}$$

$$C = -m_0 \times n - k \times \frac{e^{(n-N_k)\times\delta} - e^{-N_k\times\delta}}{\delta} + C_{100} \tag{24}$$

The capacity of the first 100 cycles is used to determine the initial capacity fade rate. Afterwards, the unknown parameters in Equation (24) are optimized using the nonlinear least squares method, so that Equation (24) can better fit the predicted 10 future capacity fade curve sampling points from the TM-Seq2Seq model.

4. Results and Discussion

In this section, the predicting results of our model and the fitting results of the capacity fade curve are presented. The prediction errors of different models are then compared.

To comprehensively evaluate the prediction performance of capacity fade curves, root mean square error (RMSE), mean absolute error (MAE), and mean absolute percentage error (MAPE) are utilized as error quantification metrics to assess the accuracy of the estimation results. Additionally, the predicted label $\boldsymbol{y}_i = (y_{i1}, y_{i2}, \cdots, y_{i10})$ reflects the rate and trend of capacity fade, and y_{i10} represents the battery's end of life (EOL). The prediction result for EOL is crucial for formulating maintenance and recycling plans for the battery. Therefore, the prediction errors of the sampling sequence and EOL are separately calculated as shown below:

$$RMSE = \sqrt{\frac{1}{10 \times n}\sum_{i=1}^{n}\sum_{j=1}^{10}\left(\hat{y}_{ij} - y_{ij}\right)^2} \tag{25}$$

$$RMSE_{EOL} = \sqrt{\frac{1}{n}\sum_{j=1}^{n}\left(\hat{y}_{i10} - y_{i10}\right)^2} \tag{26}$$

$$MAE = \frac{1}{10 \times n}\sum_{i=1}^{n}\sum_{j=1}^{10}\left|\hat{y}_{ij} - y_{ij}\right| \tag{27}$$

$$MAE_{EOL} = \frac{1}{n}\sum_{i=1}^{n}\left|\hat{y}_{i10} - y_{i10}\right| \tag{28}$$

$$MAPE = \frac{100\%}{10 \times n}\sum_{i=1}^{n}\sum_{j=1}^{10}\left|\frac{\hat{y}_{ij} - y_{ij}}{y_{ij}}\right| \tag{29}$$

$$MAPE_{EOL} = \frac{100\%}{n}\sum_{j=1}^{10}\left|\frac{\hat{y}_{i10} - y_{i10}}{y_{i10}}\right| \tag{30}$$

where n represents the total number of test samples. RMSE considers the square of estimation errors, thus penalizing larger errors more heavily and better reflecting the differences between model estimations and actual observations. MAE calculates the absolute value of estimation errors, directly measuring the average difference between estimated values and observed values. MAPE converts estimation errors into percentage form, intuitively reflecting the magnitude of estimation errors relative to observed values, facilitating the evaluation of model performance on different datasets.

To demonstrate the effectiveness of the model, comparative experiments were conducted against other popular machine learning architectures. We used the programming software 'Python 3.9' to build, train, and test all models. The participating models for comparison are as follows:

1. Seq2Seq: The Seq2Seq model consists of an encoder and a decoder. The encoder transforms the input sequence into a fixed-length vector, which is then used by the decoder to generate the output sequence. This model has demonstrated excellent performance in tasks such as predicting capacity fade curves [15].
2. CNN: The CNN model captures local features through convolution operations and reduces the number of parameters and computational complexity through pooling operations. It is also proficient in learning time series data. This model has been widely applied to tasks such as battery EOL prediction [18] and capacity fade curve prediction [22,23].
3. CNN-BI-LSTM: BI-LSTM is a type of recurrent neural network model used for natural language processing tasks such as named entity recognition and sentiment analysis. The model comprises both forward and backward LSTM layers, allowing it to utilize contextual information at each time step and better capture long-term dependencies. This model has been applied to tasks of battery SOH prediction [31].
4. SE-CNN-LSTM: SE-CNN-LSTM is a model that combines Squeeze-and-Excitation (SE) module, Convolutional Neural Network (CNN), and Long Short-Term Memory (LSTM). It leverages the SE module to enhance the feature representation of CNN, which allows the model to capture more informative features.

Each model is trained and tested under similar conditions to ensure a fair comparison. We conducted 10 experiments. In each of the 10 experiments, we randomly divided 132 sets of batteries into three groups: a training set (~60%), a validation set (~20%), and a test set (~20%). The model parameters were trained using the training set, and the parameters with the highest accuracy on the validation set were saved to ensure the model's performance on unseen data (generalization). Following this, the test data were input into the saved model to obtain prediction results. Finally, we calculated the average error from the 10 experiments for model comparison. The comparative results are presented in Table 1.

Table 1. Prediction errors of different models on the testing set.

Model	MAE	MAPE	RMSE
Seq2Seq	83.326	11.768%	109.921
CNN	99.730	15.619%	126.192
CNN-BI-LSTM	85.099	13.749%	105.192
SE-CNN-LSTM	90.418	13.898%	109.803
TM-Seq2Seq	77.043	12.246%	101.336
	MAE (EOL)	MAPE (EOL)	RMSE (EOL)
Seq2Seq	99.988	11.511%	127.745
CNN	113.772	14.853%	137.954
CNN-BI-LSTM	103.319	12.642%	131.097
SE-CNN-LSTM	103.478	12.207%	125.899
TM-Seq2Seq	87.838	10.409%	114.501

To justify the rationale behind incorporating the trend matching method and SE-Net, ablation experiments were performed on SE-Net and the trend matching method. The experimental results are shown in Table 2.

In addition, to visually represent the error distribution of different batteries, a distribution chart of the MAPE of all batteries is plotted, as shown in Figure 6. The width of each bar represents the corresponding error interval, and the height represents the number of batteries in that error interval. Figure 7 shows the predicted results of the EOL.

Table 2. Results comparison for ablation experiment.

SE-net Trend Matching	x x	√ x	x √	√ √
MAE	88.502	81.440	85.089	77.043
MAPE	13.171%	13.088%	13.330%	12.246%
RMSE	114.178	108.281	122.588	101.336
MAE (EOL)	107.513	91.506	97.728	87.838
MAPE (EOL)	12.757%	11.384%	12.084%	10.409%
RMSE (EOL)	135.346	112.921	137.406	114.501

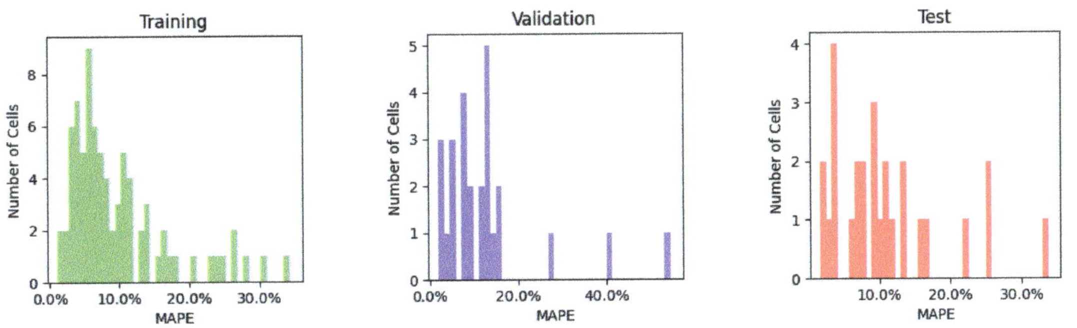

Figure 6. Histogram of error distribution for all batteries.

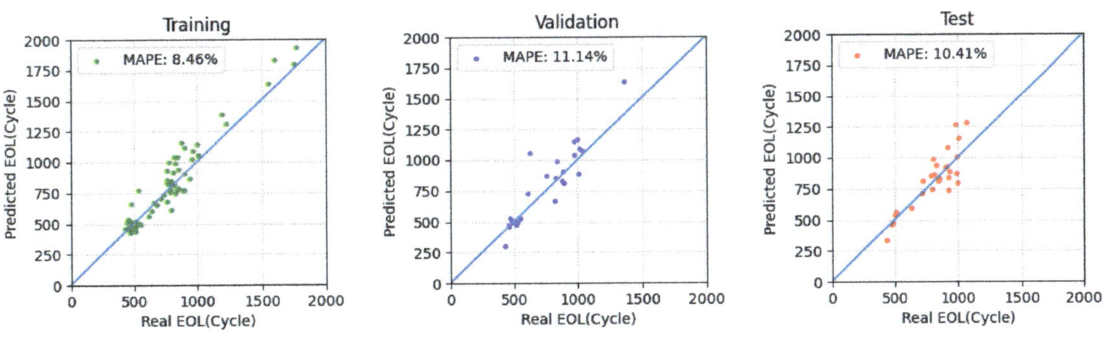

Figure 7. The predicted results of the EOL.

The researchers ranked the 27 batteries from the test set, with the first-ranked prediction being the most accurate and the predictions with lower rankings having larger errors. To showcase the fitting results of the capacity fade curve, nine batteries were selected for demonstration purposes, which are the 1th, 4th, 7th, 11th, 14th, 17th, 20th, 24th, and 27th batteries, as shown in Figure 8.

The TM-Seq2Seq model achieves the best overall performance in predicting both the capacity degradation trajectory and EOL. The ultimately predicted capacity degradation curve accurately reflects the true rate and trend of capacity fade. Additionally, incorporating both SE-net and trend matching methods improves the overall performance of the model. However, adding SE-net alone increases prediction errors. SE-net introduces channel importance weights, which enhance the model's learning capability but also increase the risk of overfitting.

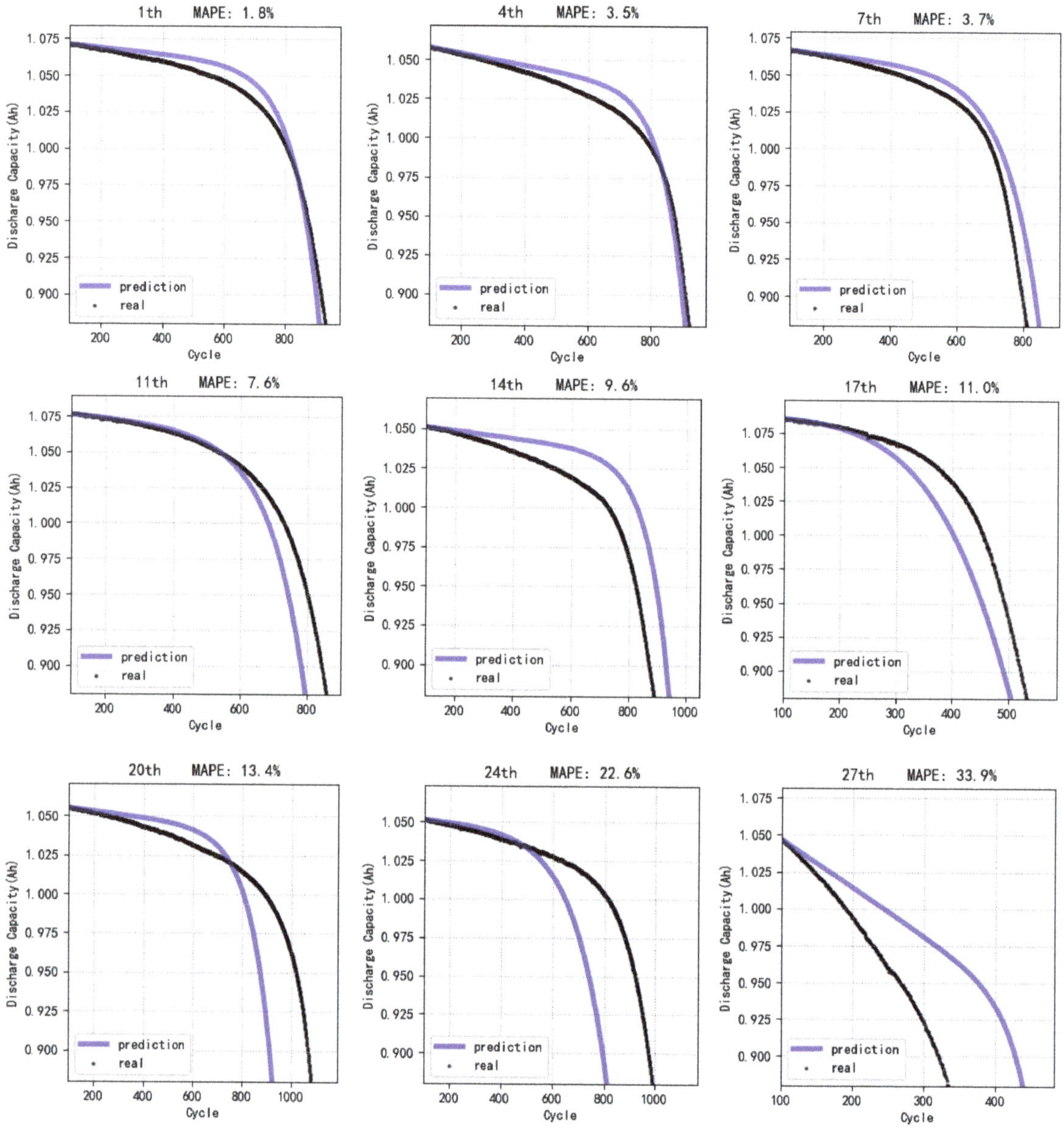

Figure 8. TM-Seq2Seq model predicted capacity fade curves for test data.

5. Conclusions

In conclusion, our study focuses on the prediction of capacity fade curves in lithium-ion batteries under different fast charging strategies. We proposed a battery capacity degradation trajectory prediction method. This method uses data from the first 100 cycles to predict the future capacity fade curve and EOL (end of life) in one-time. First, features are extracted from the discharge voltage–capacity curve. Secondly, the TM-Seq2Seq model based on CNN, SE-net, and GRU is designed. Finally, a trend matching loss function is designed based on the common characteristics of capacity fade curves. Through experimentation, the following conclusions were discovered: The SE-net enhances the model's fitting capability, allowing it to capture more intricate patterns in the data. Furthermore, the trend matching is suitable for the data conditions in this study, and it can enhance the model's performance on unknown test data. Ultimately, the TM-Seq2Seq model improves the accuracy of one-time predictions for battery capacity fade curves under various fast charging strategies.

In the future, we will further optimize this method to improve prediction performance and generalization ability. We will also extend the idea of trend feature distribution matching to other machine learning and deep learning methods to further improve the accuracy and stability of capacity fade curve prediction. The method proposed in this paper is not only applicable to battery data processing but can also be applied to similar problems in other fields, with strong versatility and scalability. This method can improve the practicality and reliability of models and is of great significance for solving practical problems.

Author Contributions: Conceptualization, X.H. and Z.D.; methodology, X.H. and Z.D.; software, Z.D.; validation, Z.D., X.H. and J.C.; formal analysis, M.R.; investigation, Z.D.; resources, X.H.; data curation, Y.S.; writing—original draft preparation, Z.D.; writing—review and editing, X.H.; visualization, Z.D.; supervision, X.H.; project administration, X.H.; funding acquisition, X.H. All authors have read and agreed to the published version of the manuscript.

Funding: This research was funded by Research Project Supported by Shanxi Scholarship Council of China, grant number 2020-041.

Data Availability Statement: Publicly available datasets were analyzed in this study. This data can be found here: 1. https://data.matr.io/1/projects/5c48dd2bc625d700019f3204, accessed on 1 January 2024. 2. https://data.matr.io/1/projects/5d80e633f405260001c0b60a, accessed on 1 January 2024.

Conflicts of Interest: The authors declare no conflicts of interest.

References

1. Chen, M.; Ma, G.; Liu, W.; Zeng, N.; Luo, X. An overview of data-driven battery health estimation technology for battery management system. *Neurocomputing* **2023**, *532*, 152–169. [CrossRef]
2. Hu, X.; Xu, L.; Lin, X.; Pecht, M. Battery Lifetime Prognostics. *Joule* **2020**, *4*, 310–346. [CrossRef]
3. Fuller, T.F.; Doyle, M.; Newman, J. Simulation and optimization of the dual lithium ion insertion cell. *J. Electrochem. Soc.* **1994**, *141*, 1. [CrossRef]
4. Doyle, M.; Fuller, T.F.; Newman, J. Modeling of galvanostatic charge and discharge of the lithium/polymer/insertion cell. *J. Electrochem. Soc.* **1993**, *140*, 1526–1533. [CrossRef]
5. Santhanagopalan, S.; Guo, Q.; Ramadass, P.; White, R.E. Review of models for predicting the cycling performance of lithium ion batteries. *J. Power Sources* **2006**, *156*, 620–628. [CrossRef]
6. Kindermann, F.M.; Keil, J.; Frank, A.; Jossen, A. A SEI modeling approach distinguishing between capacity and power fade. *J. Electrochem. Soc.* **2017**, *164*, E287–E294. [CrossRef]
7. Deshpande, R.; Verbrugge, M.; Cheng, Y.-T.; Wang, J.; Liu, P. Battery cycle life prediction with coupled chemical fade and fatigue mechanics. *J. Electrochem. Soc.* **2012**, *159*, A1730–A1738. [CrossRef]
8. Reniers, J.M.; Mulder, G.; Howey, D.A. Review and performance comparison of mechanical-chemical fade models for lithium-ion batteries. *J. Electrochem. Soc.* **2019**, *166*, A3189–A3200. [CrossRef]
9. Saha, B.; Goebel, K.; Poll, S.; Christophersen, J. Prognostics methods for battery health monitoring using a Bayesian framework. *IEEE Trans. Instrum. Meas.* **2009**, *58*, 291–296. [CrossRef]
10. He, W.; Williard, N.; Osterman, M.; Pecht, M. Prognostics of lithium-ion batteries based on Dempster-Shafer theory and the Bayesian Monte Carlo method. *J. Power Sources* **2011**, *196*, 10314–10321. [CrossRef]
11. Li, P.; Zhang, Z.; Xiong, Q.; Ding, B.; Hou, J.; Luo, D.; Rong, Y.; Li, S. State-of-health estimation and remaining useful life prediction for the lithium-ion battery based on a variant long short term memory neural network. *J. Power Sources* **2020**, *459*, 228069. [CrossRef]
12. Li, X.; Zhang, L.; Wang, Z.; Dong, P. Remaining useful life prediction for lithium-ion batteries based on a hybrid model combining the long short-term memory and Elman neural networks. *J. Energy Storage* **2019**, *21*, 510–518. [CrossRef]
13. Wang, F.; Zhao, Z.; Ren, J.; Zhai, Z.; Wang, S.; Chen, X. A transferable lithium-ion battery remaining useful life prediction method from cycle-consistency of fade trend. *J. Power Sources* **2022**, *525*, 231027.
14. Zhang, Y.; Xiong, R.; He, H.; Pecht, M.G. Long short-term memory recurrent neural network for remaining useful life prediction of lithium-ion batteries. *IEEE Trans. Veh. Technol.* **2018**, *67*, 5695–5705. [CrossRef]

15. Li, W.; Sengupta, N.; Dechent, P.; Howe, D.; Annaswamy, A.; Sauer, D.U. One-shot battery fade trajectory prediction with deep learning. *J. Power Sources* **2021**, *506*, 230024. [CrossRef]
16. Severson, K.A.; Attia, P.M.; Jin, N.; Perkins, N.; Jiang, B.; Yang, Z.; Chen, M.H.; Aykol, M.; Herring, P.K.; Fraggedakis, D.; et al. Data-driven prediction of battery cycle life before capacity fade. *Nat. Energy* **2019**, *4*, 383–391. [CrossRef]
17. Ma, Y.; Wu, L.; Guan, Y.; Peng, Z. The capacity estimation and cycle life prediction of lithium-ion batteries using a new broad extreme learning machine approach. *J. Power Sources* **2020**, *476*, 228581. [CrossRef]
18. Shen, S.; Nemani, V.; Liu, J.; Hu, C.; Wang, Z. A hybrid machine learning model for battery cycle life prediction with early cycle data. In Proceedings of the 2020 IEEE Transportation Electrification Conference and Expo (ITEC), Chicago, IL, USA, 23–26 June 2020; pp. 181–184.
19. Hong, J.; Lee, D.; Jeong, E.R.; Yi, Y. Towards the swift prediction of the remaining useful life of lithium-ion batteries with end-to-end deep learning. *Appl. Energy* **2020**, *278*, 115646. [CrossRef]
20. Fermín-Cueto, P.; McTurk, E.; Allerhand, M.; Medina-Lopez, E.; Anjos, M.F.; Sylvester, J.; Reis, G.D. Identification and machine learning prediction of knee point and knee-onset in capacity fade curve of lithium-ion cells. *Energy AI* **2020**, *1*, 100006. [CrossRef]
21. Liu, J.; Thelen, A.; Hu, C.; Yang, X.G. An end-to-end learning framework for early prediction of battery capacity trajectory. In Proceedings of the Annual Conference of the PHM Society, Nashville, TN, USA, 29 November–20 December 2021; Volume 13.
22. Saxena, S.; Ward, L.; Kubal, J.; Lu, W.; Babinec, S.; Paulson, N. A convolutional neural network model for battery capacity fade curve prediction using early life data. *J. Power Sources* **2022**, *542*, 231736. [CrossRef]
23. Strange, C.; Reis, G.D. Prediction of future capacity and internal resistance of Li-ion cells from one cycle of input data. *Energy AI* **2021**, *5*, 100097. [CrossRef]
24. Herring, P.; Gopal, C.B.; Aykol, M.; Montoya, J.H.; Anapolsky, A.; Attia, P.M.; Gent, W.; Hummelshoj, J.S.; Hung, L.; Kwon, H.; et al. A Python library for battery evaluation and early prediction. *Software* **2020**, *11*, 100506. [CrossRef]
25. Attia, P.M.; Grover, A.; Jin, N.; Severson, K.A.; Markov, T.M.; Liao, Y.; Chen, M.H.; Cheong, B.; Perkins, N.; Yang, Z.; et al. Closed-loop optimization of fast-charging protocols for batteries with machine learning. *Nature* **2020**, *578*, 397–402. [CrossRef] [PubMed]
26. Sutskever, I.; Vinyals, O.; Le, Q.V. Sequence to Sequence Learning with Neural Networks. In Proceedings of the 2014 Proceedings of the 27th International Conference on Neural Information Processing Systems (NIPS), Montreal, QC, Canada, 8–13 December 2014; pp. 3104–3112.
27. Hu, J.; Shen, L.; Sun, G. Squeeze-and-Excitation Networks. In Proceedings of the 2018 IEEECVF Conference on Computer Vision and Pattern Recognition (CVPR), Salt Lake City, UT, USA, 18–23 June 2018.
28. Song, Y.; Li, L.; Peng, Y.; Liu, D. Lithium-Ion Battery Remaining Useful Life Prediction Based on GRU-RNN. In Proceedings of the 2018 12th International Conference on Reliability, Maintainability, and Safety (ICRMS), Shanghai, China, 17–19 October 2018; pp. 317–322.
29. Zhou, K.; Liu, Z.; Qiao, Y.; Xiang, T.; Loy, C.C. Generalizing to Unseen Domains: A Survey on Domain Generalization. *IEEE Trans. Pattern Anal. Mach. Intell.* **2022**, *35*, 4396–4415.
30. Du, Y.; Wang, J.; Feng, W.; Pan, S.; Qin, T.; Xu, R.; Wang, C. AdaRNN: Adaptive Learning and Forecasting for Time Series. In Proceedings of the 2021 Proceedings of the 30th ACM International Conference on Information & Knowledge Management (CIKM), Gold Coast, QD, Australia, 1–5 November 2021; pp. 402–414.
31. Yan, B.; Zheng, W.J.; Tang, D.Y.; Laili, Y.J.; Xing, Y. A knowledge-constrained CNN-BiLSTM model for lithium-ion batteries state-of-charge estimation. *Microelectron. Reliab.* **2023**, *150*, 115112. [CrossRef]

batteries

MDPI

Article

Fast Impedance Spectrum Construction for Lithium-Ion Batteries Using a Multi-Density Clustering Algorithm

Ling Zhu [1], Jichang Peng [1], Jinhao Meng [2,*], Chenghao Sun [3], Lei Cai [4] and Zhizhu Qu [3]

1 Smart Grid Research Institute, Nanjing Institute of Technology, Nanjing 211167, China; y00450210708@njit.edu.cn (L.Z.); pengjichang@njit.edu.cn (J.P.)
2 School of Electrical Engineering, Xi'an Jiaotong University, Xi'an 710049, China
3 Faculty of Computer Science and Engineering, Xi'an University of Technology, Xi'an 710048, China; sunchenghao@njnu.edu.cn (C.S.); 21220524@njnu.edu.cn (Z.Q.)
4 School of Electrical and Automation Engineering, Nanjing Normal University, Nanjing 210023, China; cailei@xaut.edu.cn
* Correspondence: jinhao@xjtu.edu.cn

Abstract: Effectively extracting a lithium-ion battery's impedance is of great importance for various onboard applications, which requires consideration of both the time consumption and accuracy of the measurement process. Although the pseudorandom binary sequence (PRBS) excitation signal can inject the superposition frequencies with high time efficiency and an easily implementable device, processing the data of the battery's impedance measurement is still a challenge at present. This study proposes a fast impedance spectrum construction method for lithium-ion batteries, where a multi-density clustering algorithm was designed to effectively extract the useful impedance after PRBS injection. According to the distribution properties of the measurement points by PRBS, a density-based spatial clustering of applications with noise (DBSCAN) was used for processing the data of the lithium-ion battery's impedance. The two key parameters of the DBSCAN were adjusted by a delicate workflow according to the frequency range. The validation of the proposed method was proved on a 3 Ah lithium-ion battery under nine different test conditions, considering both the SOC and temperature variations.

Keywords: lithium-ion battery; pseudorandom sequence; electrochemical impedance spectroscopy; signal denoising; density clustering

1. Introduction

In recent years, green low-carbon development has gradually become the consensus and development goal of our future society. Under the demand of the low-carbon concept, a high proportion of intermittent and stochastic renewable energy sources merging into the grid will bring new challenges to the stability of the grid's operation. The energy storage technology represented by lithium-ion batteries can cope with the challenges and provide technologies and solutions for power regulation and supporting the stability of the grid [1–3]. A lithium-ion battery has high energy density and a fast response speed, which can meet the scheduling needs of the grid. However, a lithium-ion battery based on an energy storage system needs to be combined with the states to control the charging and discharging process to avoid overcharging, over-discharging, and other safety issues. Thus, an efficient battery energy management system (BMS) is the key to the reliable operation of a battery energy storage system [4–6].

A BMS contains several functions, such as signal measurement, cell equalization, state estimation, and protective fault warnings. The signal measurement module aims to measure the voltage and temperature in real time through the analog front-end (AFE) and monitors the batteries' external performance [7,8]. The equalization circuit enables a consistent control of individual cells by transferring energy. The function of the state estimation is to capture the internal states of the battery, including state of charge (SOC) and state of health (SOH), thus, providing data support for controlling, operating, and maintaining the battery energy storage system [9,10]. The protection and warning function combines external measurement and the internal state to realize the regulation of and protection against abnormal working conditions during the battery's operation, as well as warnings of safety-related failures [11–13]. Current BMS systems rely on the detection of basic electrothermal information such as voltage, current, and temperature; however, it is difficult to effectively obtain the changes in the internal electrochemical states of the battery. To realize the accurate estimation of SOC and SOH, there is an urgent need to carry out research on methodologies of detecting the battery's electrochemical state for onboard applications [14,15].

Electrochemical impedance spectroscopy (EIS) is a technique for obtaining changes in the impedance of a battery in the broadband frequency range and realizing the observation of electrochemical reactions such as lithium ions' kinetics, charge transfer, and diffusion process, and thus can reflect the battery's state. The impedance spectrum provides a new characterization method for sensing the internal states of a battery, which has gained extensive attention and been the focus of much research in recent years [16,17]. The authors of [18] connected the trend of the impedance spectral curve to the degradation process of 18650 lithium-ion batteries and explored the impedance characteristics corresponding to the batteries' SOC and SOH. It was concluded that the ohmic resistance could reflect the batteries' SOH and the transfer resistance was closely related to the variation in SOC. The authors of [19] used EIS to establish a frequency domain model and then chose the parameters of the model as the features to achieve SOC estimation using light computing machine learning algorithms. The authors of [20] used the imaginary parts of the mid-frequency band in the EIS, which were believed to have no obvious dependence on the temperature, as a feature for predicting the battery's remaining useful life (RUL) with a one-dimensional convolutional neural network-self-attention mechanism, and achieved an accurate RUL of lithium-ion batteries.

As a non-invasive technology for estimating a battery's, EIS has received increasing attention. However, a battery's EIS measurement uses electrochemical workstations applying the sine sweeping method. It has been realized that the sine sweeping method is time-consuming and needs expensive test equipment, and is mostly used for laboratory measurements. The realization of fast online measurement of impedance is a current research hotspot for battery system applications.

The authors of [21] proposed a multi-sine wave aggregation detection method, where multi-frequency sinusoidal superposition was performed at the same time to realize the detection of impedance over the entire frequency range. Fast detection of the impedance spectrum within the 0.1 Hz–1 kHz frequency interval was achieved using a 20 s step excitation [22]. The pseudorandom binary sequence (PRBS), as an easily generated excitation signal with good power consistency, has received increasing attention in the rapid detection of batteries' impedance. The authors of [23,24] investigated the spectral characterization of the PRBS signal and the extraction method of the cell's impedance. The authors of [25,26] proposed an optimized three-level sequence to improve the power spectral density of conventional PRBS excitation, which was suitable for measuring batteries' impedance. The authors of [27] proposed a novel hybrid pseudorandom sequence (HPRS) to boost the signal power in the low-frequency region.

The impedance measurement method based on PRBS excitation is characterized by a fast measurement speed and high-power consistency. Due to the limitation in the amplitude of the signal, the power density at each frequency of the measurement interval is generally low and susceptible to interference from the measurement noise, which leads to large impedance errors. Therefore, filtering and processing the original data from the impedance calculation is an important challenge for the fast measurement of batteries' impedance. The authors of [28] proposed a new measurement evaluation procedure (MEP) and a two-dimensional Gaussian filter (2-DGF) to improve the performance of PRS. In [29,30], the moving average filter (MAF) was used to smooth the impedance curve and reduce the measurement error, which was suitable for suppressing the noise in the middle and low-frequency ranges. The authors of [31,32] proposed an impedance denoising and filtering method using the power spectral density and frequency characteristics of PRBS to accurately extract the impedance in the full frequency interval [0.21 Hz, 3.5 kHz]. It was noticed that the impedance extraction method based on MAF had a large calculation error because it failed to effectively eliminate the measurement noise. In addition, the processing based on features of fusion such as frequency and power spectra increased the dimensionality and computational effort of data processing in extraction of the impedance.

This study proposes a filtering and denoising method using impedance data's sparsity according to the inherent properties of PRBS excitation. By characterizing the distribution of the original impedance data, the noise, and useful impedance were identified to effective extraction of the impedance spectrum. The main contributions of this study are as follows:

(1) A density-based spatial clustering of applications with a noise-based (DBSCAN-based) impedance data processing method is proposed to utilize the distribution properties of the measurement points by PRBS excitation.
(2) A parameter setting workflow was designed to properly cluster the data points of the impedance measured by DBSCAN, where the key parameters, that is, the neighborhood radius and the minimum number of points within the specific neighborhood, can be adjusted according to the frequency ranges.
(3) The fast measurement of a battery's impedance by integrating the PRBS and the data processing method was verified at various temperatures and SOCs on a 3 Ah lithium-ion battery.

This article is divided into five parts. Section 2 introduces the principle of fast measurement of the impedance spectrum. Section 3 describes the extraction method for the multi-density distribution EIS data of lithium batteries. Section 4 introduces the experimental validation. Section 5 summarizes the above work.

2. The Principle of Rapid Measurement of a Battery's Impedance Spectrum

A highly aggregated carrier, PRBS, was chosen as the excitation signal to realize the extraction of the full frequency band impedance spectrum of a lithium-ion battery in a short period. In this section, the principle of PRBS generation and the method for calculating the battery's impedance spectrum are introduced.

2.1. PRBS Generation

In this study, the maximum length sequence (MLS) (a typical PRBS) was used as the excitation signal for measuring the battery's impedance. The MLS can be generated offline by shifts and different or simultaneous operations, and its algorithmic structure is shown in Figure 1.

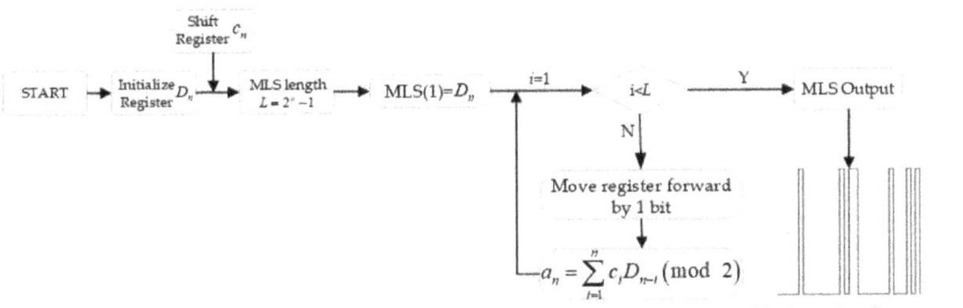

Figure 1. The generation of MLS.

In Figure 1, D_n represents the states of the n shift registers at a given moment, c_n denotes the feedback coefficients of the shift register, and L is the total length of the sequence. The first bit of the MLS is the n-th bit of the initialization register; next, the adder uses the addition of Module 2.

The general form of the linear feedback logic expression is shown in Equation (1):

$$a_n = c_1 D_{n-1} \oplus c_2 D_{n-2} \oplus \cdots \oplus c_n D_0 = \sum_{i=1}^{n} c_i D_{n-i} (\text{mod } 2) \tag{1}$$

According to the MLS generation flowchart above, the data bits in the shift register c_n are shifted one bit to the right register from D_{n-1} to D_{n-2} in each clock cycle f_{gen} to produce a new bit a_n. The complete MLS sequence is obtained after $L \times f_{gen}$ cycles, where f_{gen} is the clock period of the algorithm and L is the total length of the sequence. Moreover, f_{gen} limits the upper range of the measurable frequency for the MLS to $f_{max} = 0.45 f_{gen}$. The register order l determines the length of the MLS as $L = 2^l - 1$ and L limits the resolution of the measured frequency to $f_{min} = f_{gen}/L$.

2.2. Calculation of the Battery's Impedance

During testing a battery's impedance, it is necessary to set the parameters f_{gen} and generate the MLS sequence offline according to the measurement bandwidth $[f_{\min}, f_{\max}]$ for the battery to be tested. Afterward, the MLS sequence is injected into the lithium-ion battery as an excitation, and the battery's impedance in the whole frequency domain is calculated by collecting the voltage and current measurements. Thus, a spectral curve of the impedance is obtained. The whole process is shown in Figure 2.

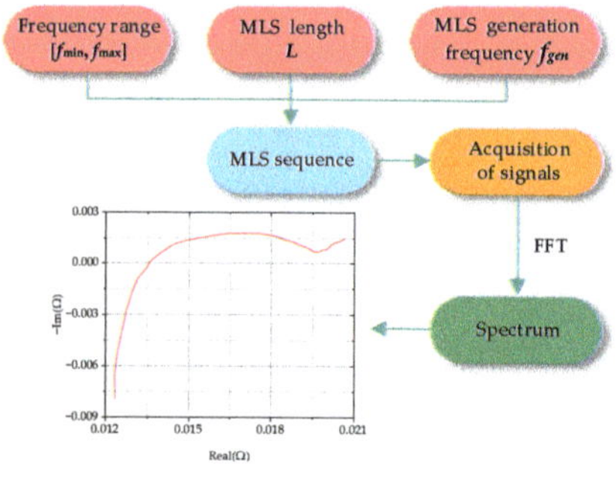

Figure 2. Impedance spectrum using the FFT transform.

The exact calculation process is as follows. The MLS is injected as an excitation signal into the lithium-ion battery, and the terminal voltage $u[n]$ and current $i[n]$ are collected. The frequency domain distribution of voltage and current $U(j\omega)$ and $I(j\omega)$ using FFT are calculated as shown in Equations (2) and (3).

$$U(j\omega) = \sum_{n=-\infty}^{+\infty} u[n] e^{-j\omega n} \tag{2}$$

$$I(j\omega) = \sum_{n=-\infty}^{+\infty} i[n] e^{-j\omega n} \tag{3}$$

The lithium-ion battery's impedance $Z(j\omega)$ can be calculated in the frequency domain as shown in Equation (4).

$$Z(j\omega) = U(j\omega)/I(j\omega) \tag{4}$$

$$\omega_i = 2\pi f_i \quad i = 1, 2, \ldots, N \tag{5}$$

Through the above steps, the impedance of the lithium-ion battery at different frequencies is sequentially obtained. The real part $Z_{\mathrm{Re}}(\omega)$ and the imaginary part $Z_{\mathrm{Im}}(\omega)$ of the impedance can be obtained as follows:

$$Z_{\mathrm{Re}}(\omega) = \mathrm{Re}[Z(j\omega)] \tag{6}$$

$$Z_{\mathrm{Im}}(\omega) = \mathrm{Im}[Z(j\omega)] \tag{7}$$

3. Extraction of the Lithium-Ion Battery's EIS Using Multi-Density Data Clustering

3.1. Characterization of the Distribution of Batteries' Impedance Based on PRBS Excitation

The EIS measurement relies on a small perturbation signal. In [28], it was found that the amplitude of the excitation signal should be less than 1 C without destroying the equilibrium state of the cell, and greater than 0.4 C to ensure a sufficient signal-to-noise ratio. In this study, the amplitude was chosen to be 0.8 C. To enable a convenient battery EIS measurement in a real application, this work used the PRBS excitation signal instead of the traditional single sine sweeping injection. Since the amplitude of the PRBS was limited, the decomposition of the PRBS's amplitude at each frequency point was even smaller compared with the sine sweeping method. In addition, the aggregated carrier class excitation method has a disadvantage in power density. The original impedance spectra of the battery calculated directly from the voltage and current responses under PRBS is shown in Figure 3. The direction of the arrow in the legend of Figure 3 is where the data density is concentrated. The measurable impedance data are chaotically dispersed at each frequency point, which increased the difficulties in extracting the useful impedance accurately and made it difficult to form an effective EIS for the battery.

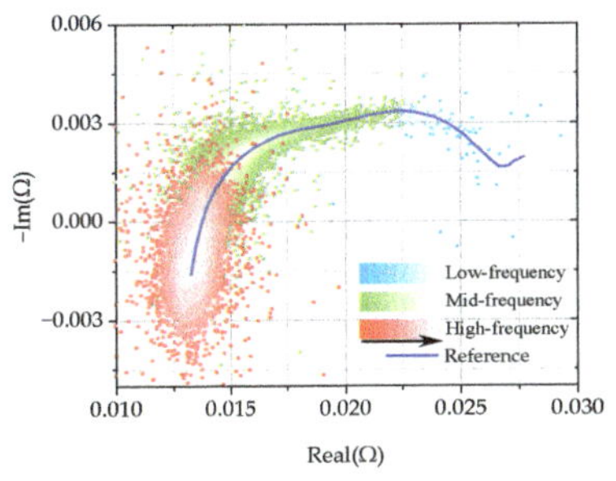

Figure 3. The original impedance spectrum obtained from the injection of the PRBS.

The low power density characteristic of PRBS excitation leads to a poor signal-to-noise ratio in the original impedance measurement results and the presence of noise interference. However, it can still be found from Figure 3 that the useful impedance and the noise had some differences in their density distribution. The effective impedance points had a relationship with the frequency variation and were arranged with high density relative to the reference (the blue line in Figure 3). Meanwhile, the noise was in a disordered and dispersed distribution. According to the basic regularity in Figure 3, we propose a new method in this work to extract the useful impedance of the battery and synchronously remove the noise with a density clustering algorithm.

3.2. Impedance Extraction Algorithm with DBSCAN

Although the battery's impedance measured by PRBS is fast and convenient, it suffers from a low signal-to-noise ratio and high difficulty in extraction. In this study, the sparse density of the original impedance measurement was characterized by DBSCAN for effective discrimination of data and noise.

3.2.1. Parameters Setting for DBSCAN

DBSCAN contains two parameters to be properly set before performing the discrimination of the impedance data's sparsity in a clustering manner, i.e., the minimum number of points in the neighborhood ***MinPts***, and the neighborhood radius ***Eps***.

(1) ***Minpts***

When processing a large-scale dataset, ***MinPts*** = 2·***D*** (***D*** is the dimension of the sample data) is usually set for DBSCAN. When the measurement data contain high noise, ***Minpts*** should be appropriately increased on the basis of 2·***D***. In this study, the ***MinPts*** was selected as 5 for clustering the battery's impedance data. This was because the battery's impedance contained two dimensions of real and imaginary parts with extensive noise.

(2) ***Eps***

The selection of the neighborhood radius ***Eps*** in DBSCAN is rather complicated, especially when dealing with large-scale and noisy battery impedance data. The parameter ***Eps*** can be selected with the aid of constructing a *k*-distance curve characterizing the sparsity of the data. This curve is made up of the distances between the data points, where the distance d_{1i} between a sample point (x_1,y_1) and all other points $(x_i,y_i \mid i \neq 1)$ in the dataset is calculated. The distance metric is based on Euclidean distance, which is calculated using Equation (8):

$$d_{1i} = \sqrt{(x_1 - x_i)^2 + (y_1 - y_i)^2} \quad i \neq 1 \tag{8}$$

After calculating the distance between any two data points in the dataset, those distances are sorted in descending order. Then, after selecting the *k* points that are closest to a sample point, the distances between these *k* points and this sample point are plotted as a *k*-distance curve. A mutation point is found at the location of the curve's inflection point. According to the elbow method, the value at the point of inflection can be set as ***Eps***.

3.2.2. Clustering Impedance Data

For the impedance data sparsity discrimination, the DBSACN needs to set the neighborhood radius ***Eps*** and the minimum number of points within the specific neighborhood ***MinPts*** to distinguish the sparsity of the data. The structure of the algorithm is shown in Figure 4.

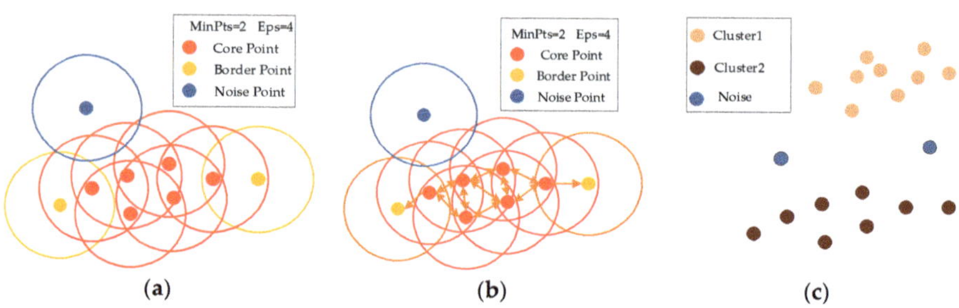

Figure 4. Schematic diagram of the clustering algorithm. The process is as follows: (**a**) selection of all core points; (**b**) constructing the neighborhood chains; (**c**) completing the classification of the clusters.

As shown in Figure 4, the formation of clusters requires three steps:

(a) Determination of the cluster cores based on clustering parameters: if a data point has more than ***MinPts*** points in its neighborhood with a radius of ***Eps***, this data point is marked as a core point. We can then find all core points in the dataset. The circles of different colors in subfigures (a) and (b) represent the neighborhood of the point with radius ***Eps***.

(b) Formation of a neighborhood chain: for each core point, find all points that are densely accessible from that core point, forming a chain of neighborhoods.

(c) Labeling the cluster classes and noise points: All points contained in a neighborhood chain are labeled as a cluster class, boundary points are assigned to the core point cluster class to which they are connected, and points that are not assigned to any cluster class are labeled as noise. The entire impedance test dataset is scanned and then labeled with noise points and several data clusters consisting of core and boundary points.

3.2.3. The Proposed Multi-Density Clustering Algorithm

After obtaining the two clustering parameters, the clustering calculation can be performed on the battery's test data. DBSCAN clusters the data according to the ***Eps*** and ***MinPts*** obtained from the calculation and divides the whole test sample set into multiple clusters. DBSCAN searches for clusters by examining the ***Eps*** neighborhood of each point in the dataset, and if the ***Eps*** neighborhood of a point p contains more points than ***MinPts*** points, a cluster is created with p as the core object. DBSCAN then iteratively aggregates the objects directly density-reachable from these core objects, a process that may involve some merging of the density-reachable clusters. The process ends when no new points are added to any cluster and a cluster class is formed, followed by repeating the process until all the data have been labeled. The steps of the algorithm are shown in Algorithm 1, where the set of core points is defined as Ω, the neighborhood of a sample point x_j is defined as the number of clustering clusters defined as k, the set of unvisited sample points is defined as Γ, the current set of unvisited sample points is defined as Γ_{old}, a queue is defined as Q, any of the core objects in Ω are defined as o, the first sample point in queue Q is defined as q, and a clustering cluster is defined as C_k.

Algorithm 1 The proposed multi-density clustering algorithm.

Input: Dataset: $D = \{x_1, x_2, \ldots, x_m\}$, ***Eps***: ε, ***MinPts***
$\Omega = \varnothing$
for $j = 1, 2, \ldots, m$ do
if $\left|N_\varepsilon\left(x_j\right)\right| \geq MinPts$ **then**
$\Omega = \Omega \cup \{x_j\}$
end if
end for
$k = 0$
$\Gamma = D$
while $\Omega \neq \varnothing$ **do**
$\Gamma_{old} = \Gamma$
$Q = \langle o \rangle$
$\Gamma = \Gamma \backslash \{o\}$
while $\Omega \neq \varnothing$ **do**
if $|N_\varepsilon(q)| \geq MinPts$ **then**
$\Delta = N_\varepsilon(q) \cap \Gamma$
$\Gamma = \Gamma \backslash \Delta$
end if
end while
$k = k+1$, $C_k = \Gamma_{old} \backslash \Gamma$
$\Omega = \Omega \backslash C_k$
end while
Output: Clustering result: $C = \{C_1, C_2, \ldots, C_k\}$

The clusters of data completed by clustering need to be type-discriminated to determine whether they are valid points or noisy points. In this study, a contour coefficient method was used to calculate the density of the data in the clusters after clustering. Data clusters with high density were judged as valid points, and data clusters with low density were treated as noisy points. The method of judging the data cluster's type is as follows

$$S(i) = \frac{b(i) - a(i)}{\max(a(i), b(i))} \tag{9}$$

where $S(i)$ represents the contour coefficient, $a(i)$ denotes the average distance from a sample point to other sample points in its cluster class, and $b(i)$ denotes the average distance of a sample point to all sample points of its nearest cluster class.

The mean value of the contour coefficients within the data clusters close to 1 ($S(i) \geq 0.9$) can be recognized as valid signals, and the rest of the clusters are the noise clusters to be eliminated from the dataset.

4. Experimental Verification

4.1. Experimental Platform

In this section, an experimental test platform for the battery's impedance spectrum was built, as shown in Figure 5. The experimental platform included a controller, an acquisition module, and a power module. First, an MLS was generated with a length of 32,767 bits using a 15-bit shift register. Second, this signal was injected into the battery via an external bipolar power supply, that is, the PBZ20-20, which was made by KIKUSUI, Yamanashi, Japan. Next, an instrumentation amplifier performed a differential operation on the signals from the two terminals of the cell to obtain the battery's response. Next, the dual-channel high-speed digital-to-analog converter used the same clock to sample the response signal and the excitation signal, and the sampling results were cached by SRAM in real time and then uploaded to the host computer. Finally, the host computer uses MatlabR2022b to perform FFT analysis on the data and calculated the spectra of the excitation and response signals to obtain the amplitude and phase characteristics of the measured impedance.

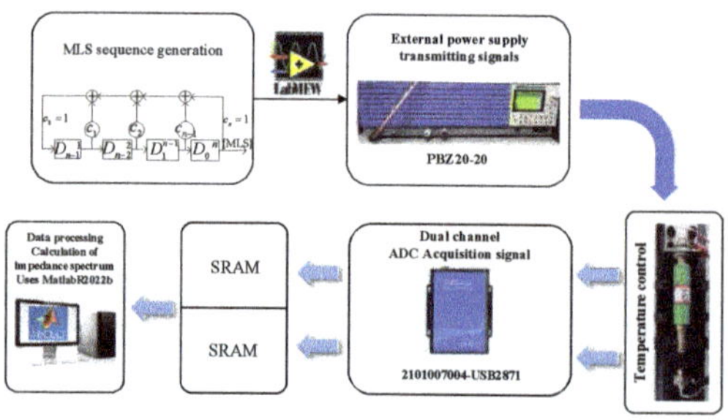

Figure 5. Construction of the experimental platform.

In this experiment, a SONY VT5C 18650 lithium-ion battery which is manufactured by SONY of Tokyo, Japan was selected as the test sample, and the battery's specifications are shown in Table 1.

Table 1. The specifications of the SONY VT5C 18650 lithium-ion battery.

Nominal Capacity	Voltage Range	Maximum Current
3000 mAh	2.0–4.25 V	30 A

4.2. Validation of the Impedance Measurement

The proposed impedance extraction algorithm utilizes the difference in the sparse density between the impedance data to remove the disordered and dispersive noise points and realize the extraction of reliable impedance measurements. In this study, the results of the battery's impedance under PRBS excitation showed some characteristics. The higher the injection frequency, the more scattered the results of the impedance calculation were.

Therefore, the whole test interval was divided into three frequency bands, namely, the low-frequency interval, the middle-frequency interval, and the high-frequency interval. There were differences in data's density at different frequency intervals, and the parameters of the corresponding impedance extraction algorithms were changed. The process of extracting the impedance is shown in Figure 6.

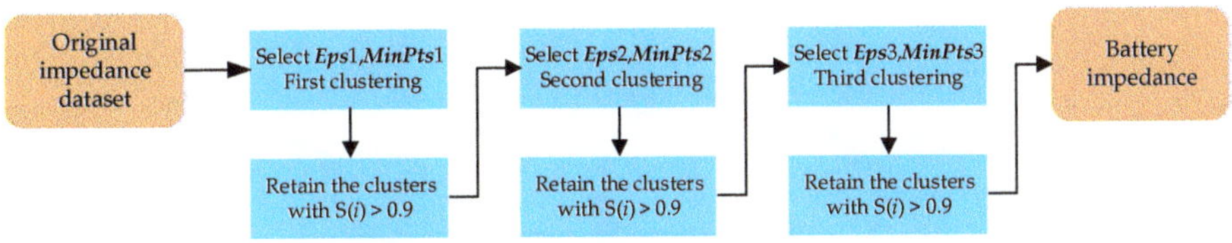

Figure 6. The process of extracting the impedance in the proposed method.

The impedance calculations in the high-frequency interval were taken as an example to verify the process of parameter selection and data processing in Section 3. As shown in Figure 7a, ***Minpts1*** was set to 5 to construct the k-distance (k = 5) curve for data in the high-frequency interval, and the curve's inflection point was selected as ***Eps1***. The first clustering was carried out using the parameters above, and the results of the data discrimination are shown in Figure 7b.

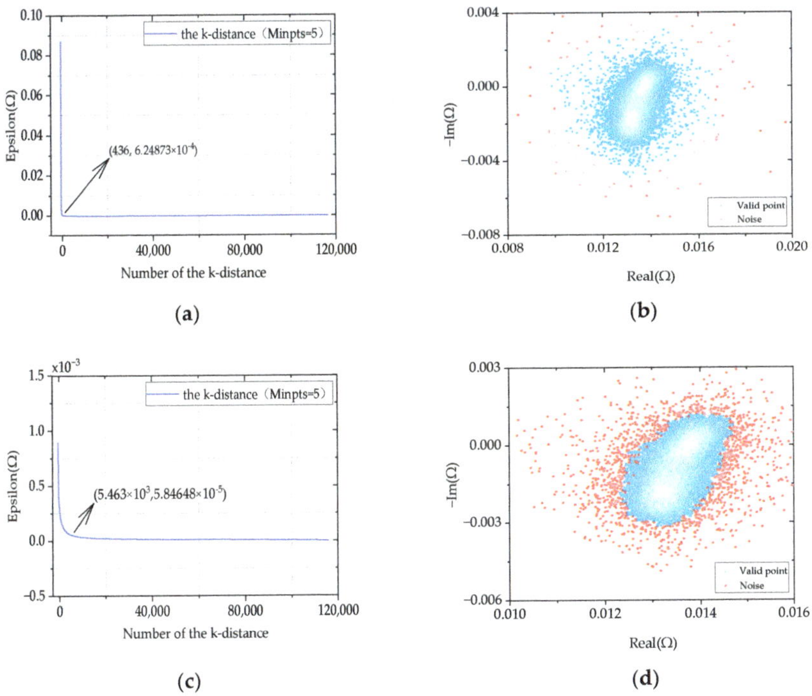

Figure 7. *Cont.*

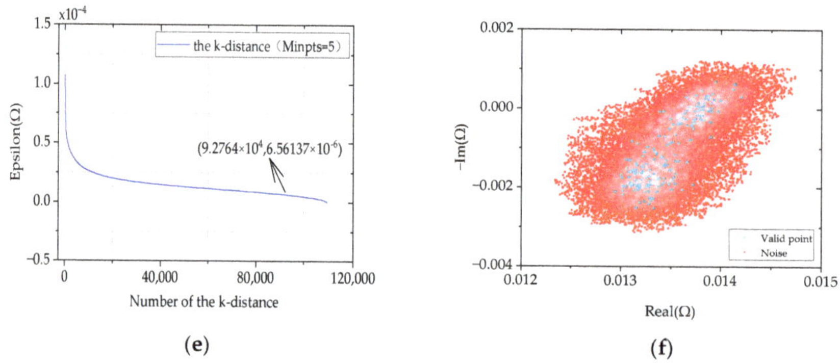

Figure 7. Process of clustering data from the high-frequency region. (**a**) The parameters selected in the first round. (**b**) The first clustering results. (**c**) The parameters selected in the second round. (**d**) The second set of clustering results. (**e**) The parameters selected for the third round. (**f**) The third set of clustering results.

From Figure 7b, it can be seen that the impedance data were classified into a peripheral data cluster with high dispersion and an inner data cluster with high aggregation after the first clustering procedure. A large amount of noise was effectively excluded by clustering even in the first round. Due to the dispersion and the large amount of data in the high-frequency interval, the ideal extraction results could not be obtained by using the clustering method only once. The useful impedance of the battery was retained for the density clustering after enough clustering had been completed. The EIS points retained after three rounds of clustering are shown in Figure 7f. The algorithm effectively filtered out a large amount of noise, and the retained data can be judged as effective test points, that is, the battery's EIS.

After processing the data in the high-frequency interval, the process above was used again to extract the effective impedance from the mid-frequency and low-frequency intervals. Due to the differences in the density characteristics of the data in the mid-frequency and low-frequency regions, it was necessary to select appropriate clustering parameters based on the distribution characteristics of the test data in different frequency intervals. For example, as the data's density decreased during the clustering process, a large value of ***MinPts*** caused two neighboring clusters to merge into a single one, causing the extraction algorithm to fail. Therefore, in the mid- and the low-frequency ranges, ***MinPts*** was reduced appropriately. Table 2 shows the parameters of clustering in the three frequency ranges.

Table 2. The parameters for clustering in the three frequency intervals.

Frequencies	*Times*	*MinPts*	*Eps*
High frequency	1	5	6.25×10^{-4}
	2		5.85×10^{-5}
	3		6.56×10^{-6}
Mid-frequency	1	4	2.65×10^{-4}
	2		6.30×10^{-5}
	3		9.64×10^{-6}
Low frequency	1	3	1.04×10^{-3}
	2		2.77×10^{-4}
	3		1.30×10^{-5}

By using the proposed method, the impedance measurements in the high-, medium-, and low-frequency intervals kept a similar number of the data points (Figure 8). It also indicated that the proposed method could effectively extract the battery's impedance after denoising.

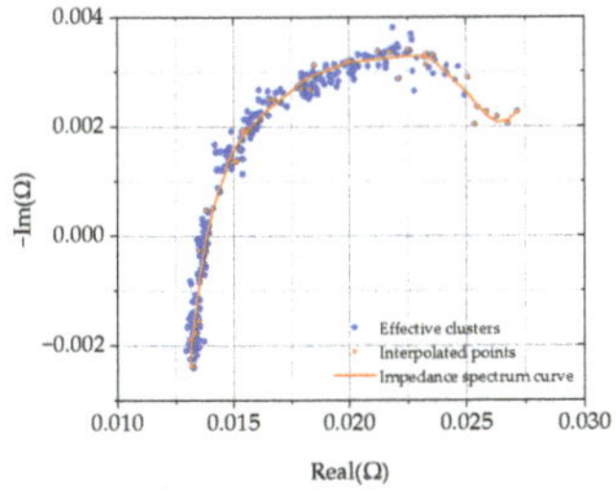

Figure 8. The impedance obtained by the proposed method.

After the effective clusters had been formed, interpolation was used to extend the relationships between the impedance data points of known frequencies to those of unknown frequencies. This resulted in a smooth impedance spectrum curve, as shown in Figure 8. The interpolated impedance spectrum offered more detailed and smoother information on the frequency response.

4.3. Validation at Different SOCs and Temperatures

In this section, the impedance curves were extracted from the sample lithium-ion battery at different SOCs and temperatures, as shown in Figure 9.

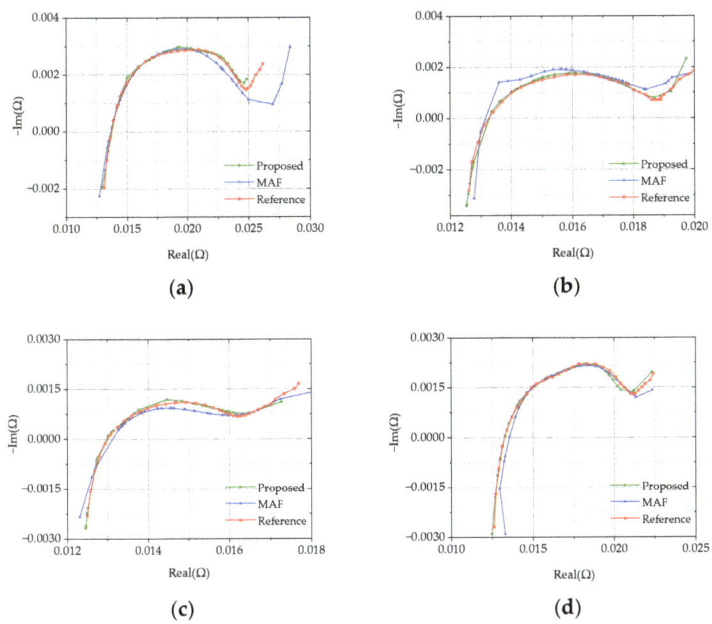

Figure 9. Lithium-ion battery's impedance under different test conditions: (**a**) T = 15 °C, SOC = 50%; (**b**) T = 25 °C, SOC = 50%; (**c**) T = 35 °C, SOC = 50%; (**d**) T = 25 °C, SOC = 80%.

The root mean square error (RMSE) and the mean absolute percentage error (MAPE) were used to evaluate the overall error for the impedance curves after data processing.

The impedance results of the lithium-ion battery in four cases are shown in Figure 9. The superiority of the proposed method was highlighted by calculating the RMSE and MAPE for the high-, medium-, and low-frequency intervals. The MAPE of the impedance in the low-frequency range obtained by the proposed method was 0.0668 when the lithium-ion battery was at 15 °C and the SOC was 50%, while the MAPE obtained by the MAF was 0.3127, which is 4.6 times higher than that of our method. As shown in Figure 9a, it can be noticed that the impedance data obtained by the MAF in the low-frequency range were far from the reference. An increase in temperature affected the impedance spectrum of the lithium-ion battery in the mid-frequency range. At 25 °C, the MAPE of the mid-frequency range measured by the MAF was twice that of the proposed method, as seen in Figure 9b. When the temperature was 35 °C, the gap between the two methods increased further, as shown in Figure 9c. The MAPE of the impedance spectrum in the mid-frequency interval measured by the MAF (0.0704) was five times higher than that of the proposed method (0.0141). Figure 9d shows the case where the Li-ion battery had SOC = 80% and T = 25 °C. It is clear that the proposed method still had significantly high accuracy, while the impedance spectrum measured by the MAF showed a large deviation in the high-frequency range. Thus, the proposed multi-density clustering method has good computational accuracy in different cases.

In contrast, traditional MAF filtering could not carefully screen the measured data points. That is because the MAF filtering process treats the useful impedance and noise data homogeneously for suppressing the noise. In the high-frequency interval, the number of noise points was much larger than the number of valid test points, so the error of MAF filtering was larger in the high-frequency interval. Due to the small number of data points in the low-frequency range, the noise had a large impact on the results from the MAF. The key for the MAF is to choose an appropriate window size and window function; however, the lithium-ion battery's impedance spectrum processed in this study had the characteristics of a non-uniform distribution of density. Thus, the MAF for processing the sparse data in the low-frequency interval was not as good as expected.

The RMSE and MAPE of the two methods in different cases were compared, as shown in Figure 10. When the lithium-ion battery was at an ambient temperature of 15 °C and 50% SOC, the difference in RMSE and MAPE between the two methods was the largest. The RMSE and MAPE of the proposed method were reduced by 85.6% and 82.3% compared with those of the MAF.

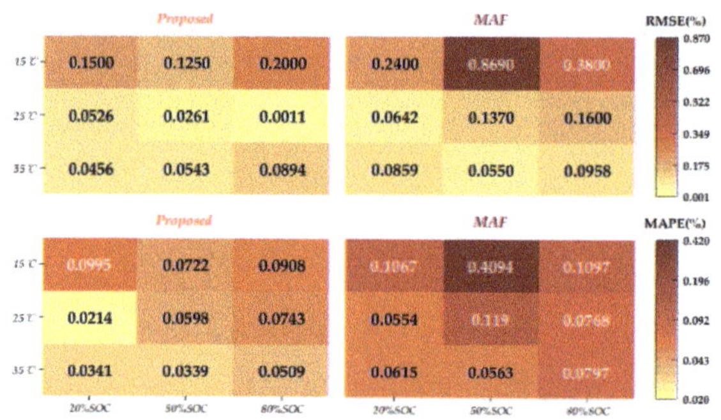

Figure 10. A comparison of the RMSE and MAPE of the two methods in nine test cases.

5. Conclusions

In this article, a denoising method for processing multi-density impedance data is proposed. The algorithm analyzes the density characteristics of the measurement data of impedance and determines two core parameters to discriminate the noisy points and valid data points by DBSCAN, realizing the cleaning of the dataset and the effective extraction of the impedance spectrum. The effectiveness and superiority of the method were verified experimentally, and the results showed that the RMSE and MAPE calculated by the method proposed were smaller than those of the MAF filtering method at different SOCs and temperatures of the lithium-ion battery. The average RMSE of the proposed method was 0.082, which was smaller than that of the MAF (0.15). The average MAPE of the method in this study was 0.06, which was smaller than that of the MAF (0.12). In addition, the curves obtained by the proposed impedance extraction algorithm were closer to the reference curve for all the validation cases. Thus, the data processing method based on multi-density clustering DBSCAN solves the problem of the low signal-to-noise ratio of PRBS in rapid measurements of EIS and improves the accuracy of the impedance measurements. The proposed method is effective and can provide accurate impedance information for the estimating the battery's status in BMS.

Author Contributions: Conceptualization, J.P. and J.M.; methodology, J.P. and L.Z.; software, J.P., L.Z. and J.M.; validation, C.S., L.C. and Z.Q.; formal analysis, J.P. and L.Z.; investigation, J.M.; writing—original draft preparation, J.P. and L.Z.; writing—review and editing, J.P. and J.M. All authors have read and agreed to the published version of the manuscript.

Funding: This work was supported by the Natural Science Foundation of China under Grant 52107229), the Key Research and Development Program of Shaanxi Province (2024GX-YBXM-442), and the Fundamental Research Funds for the Central Universities under Grant xxj032023002.

Data Availability Statement: The raw data supporting the conclusions of this article will be made available by the authors on request.

Conflicts of Interest: The authors declare no conflicts of interest.

References

1. Lawder, M.T.; Suthar, B.; Northrop, P.W.C.; De, S.; Hoff, C.M.; Leitermann, O.; Crow, M.L.; Santhanagopalan, S.; Subramanian, V.R. Battery Energy Storage System (BESS) and Battery Management System (BMS) for Grid-Scale Applications. *Proc. IEEE* **2014**, *102*, 1014–1030. [CrossRef]
2. Xiong, R.; Pan, Y.; Shen, W.; Li, H.; Sun, F. Lithium-Ion Battery Aging Mechanisms and Diagnosis Method for Automotive Applications: Recent Advances and Perspectives. *Renew. Sustain. Energy Rev.* **2020**, *131*, 110048. [CrossRef]
3. Han, X.; Ouyang, M.; Lu, L.; Li, J. Cycle Life of Commercial Lithium-Ion Batteries with Lithium Titanium Oxide Anodes in Electric Vehicles. *Energies* **2014**, *7*, 4895–4909. [CrossRef]
4. Qays, M.O.; Buswig, Y.; Hossain, M.L.; Abu-Siada, A. Recent Progress and Future Trends on State of Charge Estimation Methods to Improve Battery-Storage Efficiency: A Review. *CSEE J. Power Energy Syst.* **2019**, *8*, 105–114. [CrossRef]
5. Hao, X.; Wu, J. Online State Estimation Using Particles Filters of Lithium-Ion Polymer Battery Packs for Electric Vehicle. In Proceedings of the 2015 IEEE International Conference on Systems, Man, and Cybernetics, Hong Kong, China, 9–12 October 2015; pp. 783–788. [CrossRef]
6. Meng, J.; Cai, L.; Stroe, D.-I.; Luo, G.; Sui, X.; Teodorescu, R. Lithium-Ion Battery State-of-Health Estimation in Electric Vehicle Using Optimized Partial Charging Voltage Profiles. *Energy* **2019**, *185*, 1054–1062. [CrossRef]
7. Hannan, M.A.; Lipu, M.S.H.; Hussain, A.; Mohamed, A. A Review of Lithium-Ion Battery State of Charge Estimation and Management System in Electric Vehicle Applications: Challenges and Recommendations. *Renew. Sustain. Energy Rev.* **2017**, *78*, 834–854. [CrossRef]
8. Peng, J.; Meng, J.; Chen, D.; Liu, H.; Hao, S.; Sui, X.; Du, X. A Review of Lithium-Ion Battery Capacity Estimation Methods for Onboard Battery Management Systems: Recent Pro-Gress and Perspectives. *Batteries* **2022**, *8*, 229. [CrossRef]
9. Meng, J.; Luo, G.; Ricco, M.; Swierczynski, M.; Stroe, D.-I.; Teodorescu, R. Overview of Lithium-Ion Battery Modeling Methods for State-of-Charge Estimation in Electrical Vehicles. *Appl. Sci.* **2018**, *8*, 659. [CrossRef]
10. Li, L.; Li, Y.; Cui, W.; Chen, Z.; Wang, D.; Zhou, B.; Hong, D. A Novel Health Indicator for Online Health Estimation of Lithium-Ion Batteries Using Partial Incremental Capacity and Dynamic Voltage Warping. *J. Power Sources* **2022**, *545*, 231961. [CrossRef]
11. Feng, X.; Ouyang, M.; Liu, X.; Lu, L.; Xia, Y.; He, X. Thermal Runaway Mechanism of Lithium Ion Battery for Electric Vehicles: A Review. *Energy Storage Mater.* **2018**, *10*, 246–267. [CrossRef]
12. Han, X.; Lu, L.; Zheng, Y.; Feng, X.; Li, Z.; Li, J.; Ouyang, M. A Review on the Key Issues of the Lithium Ion Battery Degradation among the Whole Life Cycle. *eTransportation* **2019**, *1*, 100005. [CrossRef]
13. Gao, Y.; Jiang, J.; Zhang, C.; Zhang, W.; Jiang, Y. Aging Mechanisms under Different State-of-Charge Ranges and the Multi-Indicators System of State-of-Health for Lith-Ium-Ion Battery with Li(NiMnCo)O_2 Cathode. *J. Power Sources* **2018**, *400*, 641–651. [CrossRef]
14. Huang, Q.; Shen, Y.; Huang, Y.; Zhang, L.; Zhang, J. Electrochimica Acta Impedance Characteristics and Diagnoses of Automotive Lithium-Ion. *Electrochim. Acta* **2016**, *219*, 751–765. [CrossRef]

15. Wang, X.; Wei, X.; Chen, Q.; Dai, H. A Novel System for Measuring Alternating Current Impedance Spectra of Series-Connected Lithium-Ion Batteries with a High-Power Dual Active Bridge Converter and Distributed Sampling Units. *IEEE Trans. Ind. Electron.* **2021**, *68*, 7380–7390. [CrossRef]
16. Du, X.; Meng, J.; Peng, J.; Zhang, Y.; Liu, T.; Teodorescu, R. Sensorless Temperature Estimation of Lithium-Ion Battery Based on Broadband Impedance Measurements. *IEEE Trans. Power Electron.* **2022**, *37*, 10101–10105. [CrossRef]
17. Cai, J.; Zhang, L.; Wang, X.; Zhu, J.; Yuan, Y.; Wang, Y.; Wei, X.; Dai, H. Investigation of an M-Sequence Based Impedance Spectrum Acquisition Method for Lithium-Ion Batteries from the Engineering Application Perspective. *J. Energy Storage* **2023**, *59*, 106428. [CrossRef]
18. Shu, X.; Yang, W.; Yang, B.; Wei, K.; Punyawudho, K.; Liu, C. Research on EIS Characterization and Internal Morphological Changes of LIBs during Degradation Process. *Eng. Fail. Anal* **2024**, *155*, 107764. [CrossRef]
19. Buchicchio, E.; De Angelis, A.; Santoni, F.; Carbone, P.; Bianconi, F.; Smeraldi, F. Battery SOC Estimation from EIS Data Based on Machine Learning and Equivalent Circuit Model. *Energy* **2023**, *283*, 128461. [CrossRef]
20. Li, J.; Zhao, S.; Miah, M.S.; Niu, M. Remaining Useful Life Prediction of Lithium-Ion Batteries via an EIS Based Deep Learning Approach. *Energy Rep.* **2023**, *10*, 3629–3638. [CrossRef]
21. De Angelis, A.; Buchicchio, E.; Santoni, F.; Moschitta, A.; Carbone, P. Practical Broadband Measurement of Battery EIS. In Proceedings of the 2021 IEEE International Workshop on Metrology for Automotive, MetroAutomotive, Bologna, Italy, 1–2 July 2021; pp. 25–29. [CrossRef]
22. Wang, X.; Kou, Y.; Wang, B.; Jiang, Z.; Wei, X.; Dai, H. Fast Calculation of Broadband Battery Impedance Spectra Based on S Transform of Step Disturbance and Response. *IEEE Trans. Transp. Electrif.* **2022**, *7782*, 3659–3672. [CrossRef]
23. Roinila, T.; Abdollahi, H.; Santi, E. Frequency-Domain Identification Based on Pseudorandom Sequences in Analysis and Control of Dc Power Distribution Systems: A Review. *IEEE Trans. Power Electron* **2021**, *36*, 3744–3756. [CrossRef]
24. Sihvo, J.; Messo, T.; Roinila, T.; Luhtala, R. Online Internal Impedance Measurements of Li-Ion Battery Using PRBS Broadband Excitation and Fourier Techniques: Methods and Injection Design. In Proceedings of the 2018 International Power Electronics Conference, IPEC-Niigata—ECCE Asia 2018, Niigata, Japan, 20–24 May 2018; pp. 2470–2475.
25. Sihvo, J.; Messo, T.; Roinila, T.; Luhtala, R.; Stroe, D.I. Online Identification of Internal Impedance of Li-Ion Battery Cell Using Ternary-Sequence Injection. In Proceedings of the 2018 IEEE Energy Conversion Congress and Exposition, ECCE 2018, Portland, OR, USA, 23–27 September 2018; pp. 2705–2711.
26. Sihvo, J.; Stroe, D.I.; Messo, T.; Roinila, T. Fast Approach for Battery Impedance Identification Using Pseudo-Random Sequence Signals. *IEEE Trans. Power Electron.* **2020**, *35*, 2548–2557. [CrossRef]
27. Du, X.; Du, X.; Meng, J.; Meng, J.; Peng, J.; Peng, J. Hybrid Pseudorandom Sequence for Broadband Impedance Measurements of Lithium-Ion Batteries. *IEEE Trans. Ind. Electron.* **2022**, *70*, 6856–6864. [CrossRef]
28. Zhang, Y.; Du, X.; Meng, J.; Jiang, Q.; Peng, J.; Liu, T. Rapid Broadband Impedance Acquisition of Lithium-Ion Batteries Based on Measurement Evaluating and Impedance Filtering. *IEEE Trans. Transp. Electrif.* **2023**, *9*, 4888–4897. [CrossRef]
29. Sihvo, J.; Roinila, T.; Stroe, D.I. Novel Fitting Algorithm for Parametrization of Equivalent Circuit Model of Li-Ion Battery from Broadband Impedance Measurements. *IEEE Trans. Ind. Electron.* **2021**, *68*, 4916–4926. [CrossRef]
30. Gücin, T.N.; Ovacik, L. Online Impedance Measurement of Batteries Using the Cross-Correlation Technique. *IEEE Trans. Power Electron.* **2020**, *35*, 4365–4375. [CrossRef]
31. Peng, J.; Meng, J.; Du, X.; Cai, L.; Stroe, D.I. A Fast Impedance Measurement Method for Lithium-Ion Battery Using Power Spectrum Property. *IEEE Trans. Ind. Inform.* **2023**, *19*, 8253–8261. [CrossRef]
32. Meng, J.; Peng, J.; Cai, L.; Song, Z. Rapid Impedance Extraction for Lithium-Ion Battery by Integrating Power Spectrum and Frequency Property. *IEEE Trans. Ind. Electron.* **2023**, 1–10. [CrossRef]

batteries

MDPI

Article

State of Health Estimation for Lithium-Ion Battery Based on Sample Transfer Learning under Current Pulse Test

Yuanyuan Li [1], Xinrong Huang [2], Jinhao Meng [3], Kaibo Shi [4,*], Remus Teodorescu [5] and Daniel Ioan Stroe [5]

1 Electronic Information Engineering Key Laboratory of Electronic Information of State Ethnic Affairs Commission, College of Electrical Engineering, Southwest Minzu University, Chengdu 610041, China; liyuanyuanfy@163.com
2 School of Energy and Electrical Engineering, Chang'an University, Xi'an 710064, China; hxr@chd.edu.cn
3 The School of Electrical Engineering, Xi'an Jiaotong University, Xi'an 710049, China; jinhao@xjtu.edu.cn
4 School of Electronic Information and Electrical Engineering, Chengdu University, Chengdu 610106, China
5 Department of Energy Technology, Aalborg University, 9220 Aalborg, Denmark; ret@et.aau.dk (R.T.); dis@et.aau.dk (D.I.S.)
* Correspondence: skbs111@163.com

Abstract: Considering the diversity of battery data under dynamic test conditions, the stability of battery working data is affected due to the diversity of charge and discharge rates, variability of operating temperature, and randomness of the current state of charge, and the data types are multi-sourced, which increases the difficulty of estimating battery SOH based on data-driven methods. In this paper, a lithium-ion battery state of health estimation method with sample transfer learning under dynamic test conditions is proposed. Through the Tradaboost.R2 method, the weight of the source domain sample data is adjusted to complete the update of the sample data distribution. At the same time, considering the division methods of the six auxiliary and the source domain data set, aging features from different state of charge ranges are selected. It is verified that while the aging feature dimension and the demand for target domain label data are reduced, the estimation accuracy of the lithium-ion battery state of health is not affected by the initial value of the state of charge. By considering the mean absolute error, mean square error and root mean square error, the estimated error results do not exceed 1.2% on the experiment battery data, which highlights the advantages of the proposed methods.

Keywords: lithium-ion battery; state of health; transfer learning; current pulse test; aging feature

1. Introduction

Lithium-ion batteries have the advantages of high operating voltage, high energy density, long cycle life, no memory effect, and environmental friendliness. They have been widely used in portable electronic devices such as mobile phones and notebook computers, and are gradually expanding into the fields of electric vehicles and large battery energy storage. As one of the important energy storage technologies, lithium-ion batteries play an important role in changing the energy consumption structure and electrifying traditional energy sources [1–4]. However, the storage capacity of lithium-ion batteries will continue to decline as they age, which shows a decrease in power, causing the battery capacity to decay and the internal resistance to increase. In existing research, battery state of health (SOH) is usually used as a quantitative indicator to evaluate battery aging. Nevertheless, the battery SOH will be affected by many factors, such as temperature, discharge depth and state of charge (SOC), etc., which further increases the difficulty of estimating the battery SOH. Therefore, in order to avoid safety hazards caused by battery capacity fading and maximum available power reduction, it is crucial to select an appropriate estimation method to complete battery SOH identification.

Generally speaking, these methods are mainly divided into two categories: traditional estimation methods and data-driven estimation methods.

Traditional methods focus on identifying and analyzing important external characteristic parameters in lithium-ion batteries. They usually analyze the collected experimental data such as battery current, voltage, temperature, etc., and relatively directly obtain the characteristic parameters that can reflect battery degradation and estimate battery SOH. The method includes experimental analysis method [5], electrochemical model method [6] and equivalent circuit model method [7].

The experimental analysis method evaluates the battery SOH by directly measuring certain characteristic parameters of the battery, mainly including capacity measurement, ohmic internal resistance measurement and impedance measurement. Although the experimental analysis method can accurately calibrate the battery SOH, the limitations of battery experimental equipment, experimental sites, and experimental test plans will increase the difficulty of implementing this method [8]. The electrochemical model is a first-principle model that can not only accurately simulate the external characteristics of the battery, but also simulate the distribution and changes in the internal characteristics [9]. The electrochemical model can deeply describe the microscopic reactions inside the battery and has very clear physical meanings. However, the electrochemical model contains complex partial differential equations and numerous electrochemical parameters, which not only brings accuracy advantages but also greatly increases the complexity and computational load of the model. A large number of unmeasured parameters in the model affect the application of the model in battery management systems, and its ability to dynamically track changes in environmental conditions is limited. The equivalent circuit model method usually uses traditional circuit components (such as capacitors, internal resistances, voltage sources, etc.) to form a circuit network to describe the external characteristics of the battery. The equivalent circuit model has good applicability to various working states of the battery. The state equation of the model can be easily derived, and the dynamic response and attenuation behavior of the battery can be constructed by identifying the model parameters [10]. However, the estimation results under this method usually depend on the structure of the equivalent circuit, resulting in increasing model errors.

Since the aging process of a lithium-ion battery is a dynamic coupling process and its internal electrochemical mechanism is very complex, traditional battery SOH estimation methods have problems such as model complexity and limited robustness, dynamic adaptability and accuracy. They also have limited ability to describe the dynamic and static comprehensive characteristics of lithium-ion batteries, which is not conducive to practical application. The data-driven method provides another solution for estimating battery SOH. By collecting a large amount of historical data and analyzing the original data, we can only establish the relationship between battery capacity and external features from the perspective of data information, without deeply exploring the complex internal mechanism changes in lithium-ion batteries. Relying on the data-driven method's ability to autonomously learn data information, this method has received widespread attention in battery SOH estimation [11–14].

Data-driven methods include the optimization algorithm method, sample entropy method, performance characteristics method, and machine learning method [15–20]. The optimization algorithm method used is to identify the battery model parameters through a suitable algorithm and then use the model parameters to estimate the battery SOH. A multi-time joint framework for predicting the battery SOC, SOH and SOP (state of power) has been established in [18]. The sample entropy method is a useful tool for assessing the predictability of time series, while also quantifying the regularity of the data series. For example, in [21], Sun et al. have established a battery capacity model through the sample entropy of voltage and temperature sequences, and then have estimated the battery SOH. In addition, the data-driven method can also complete the battery SOH estimation by analyzing the aging characteristics related to the degradation process of lithium-ion batteries, that is, the performance characteristics method. This method is to establish a mapping relationship between these parameters and the lithium-ion battery capacity through the evolution of dominant characteristics during the degradation process [22,23]. These dominant parameters are usually named aging features. Most of the current research focuses on the aging features that are obtained in a relatively stable working environment with a single working mode. The aging features obtained in this way are time-consuming and unfavorable for practical development. Machine learning only relies on battery historical data and uses advanced algorithms to train complex non-linear degradation behaviors of battery to achieve accurate estimation of the battery SOH [24–28]. Commonly used methods include neural networks (NN), support vector machines (SVM), Monte Carlo, Gaussian process regression (GPR) and long short-term memory networks (LSTM), etc. Although the data-driven method is easy to implement, the implementability of this method is based on sufficient aging data, and it usually takes several months or even longer to obtain aging data. That is to say, there are several issues with these methods, which have to be overcome: (1) Most references that extract the battery feature need to consider the battery's entire working condition, which increases the computing burden and lab force. (2) Different working conditions will affect the performance of the lithium-ion battery in reality, but most research has extracted features from one single stable working condition (e.g., 1 C charging/discharging rate), and the performance of these features under dynamic working conditions is doubtful.

In summary, considering that in actual scenarios, batteries usually work in complex and changeable environments, it is often necessary to obtain useful battery aging features in a very short time to perform battery SOH calibration. At the same time, considering the practical factors of the high cost of obtaining label data in the battery field, the insufficient amount of data is directly related to the effect of data information analysis and mining and affects the accuracy of the battery SOH estimation. Therefore, this paper considers how to obtain effective battery data information from the current pulse test stage, extract battery aging features in a short time, and complete accurate battery SOH estimation under small sample data.

Based on the above analysis, this paper addresses the issues discussed above from three aspects. First, the response voltage curve characteristics of the pulse test current at four rates are analyzed, and the aging features in a very short time that can reflect the battery capacity decay are extracted. Then, the proposed method in this paper is verified in two cases. The first case is based on the complete current pulse test data and considers using the aging features under all SOC intervals as the battery SOH estimation model input while verifying the universality of the proposed method in this paper. The second case is to assume that the current pulse test process is incomplete while ensuring that the auxiliary data and the source domain data set come from aging feature data in different SOC intervals. Six ways of dividing the auxiliary and the source domain data set are considered to verify the mapping relationship for battery SOH estimation model of the proposed method under different SOC intervals is not affected. Finally, it is also verified that the proposed method can still ensure the accuracy of lithium-ion battery SOH estimation in the above two application modes, whether in a single-temperature operating mode or in different temperature operating modes.

In detail, the main contributions of this paper are summarized as follows:

(1) This paper analyzes the voltage response curve characteristics under pulse current, and aging features are extracted over a very short time, which provides a basis for establishing the nonlinear relationship between the battery SOH and the aging feature.
(2) In the case of inconsistent initial SOC values, the proposed method can still ensure that the estimation accuracy of the battery SOH under different SOC interval is not affected.
(3) The proposed method in this paper can still ensure the battery SOH estimation accuracy while reducing the model feature dimension and the demand for target domain label data.

This paper is organized as follows: Section 2 introduces the method of extracting the aging feature. The related algorithms Tradaboost.R2 are introduced in Section 3. It also still introduces the battery SOH estimation under the complete SOC range. Battery SOH estimations under different SOC ranges are presented in Section 4. The conclusions and future work are given in Section 5. The specific framework for the proposed method is shown in Figure 1.

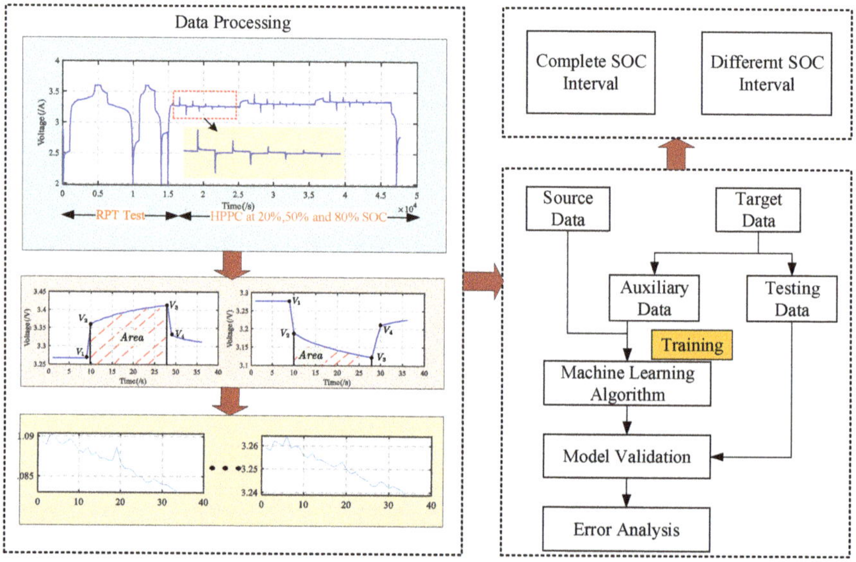

Figure 1. Battery state of health monitoring framework.

2. Aging Feature under Pulse Current

In this work, lithium iron phosphate battery cells (LFP/C) are considered, while Table 1 describing the cell's basic electric parameters. Five LFP/C cells are aged at two different temperatures, i.e., 35 °C and 25 °C, which is divided into two modes and summarised in Table 1. The capacity variations in the five LFP/C cells are presented in Figure 2. During the non-constant charging and discharging stage, three SOC and pulse currents at four rates are considered. The whole battery testing process are shown in Figure 3. First, the battery is charged to SOC = 20%, SOC = 50% and SOC = 80% with a constant current of 1 C rate. And in each SOC stage, pulse charge and discharge currents of 4 C, 2 C, 1 C and 0.5 C are applied to the battery. A negative current indicates a charging process and a positive current indicates a discharging process. Figure 4 shows the current changes in the battery under the pulse test.

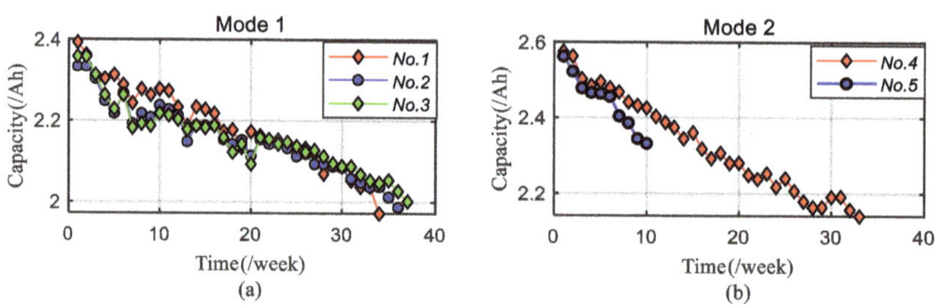

Figure 2. The capacity variation in the five LFP/C cells based on different mode.

Table 1. The working mode condition for battery data.

Test Mode	LFP/C Battery	Temperature	Average SOC
B1	No.1, No.2, No.3	35 °C	50%
B2	No.4, No.5	25 °C	50% (arbitrary)

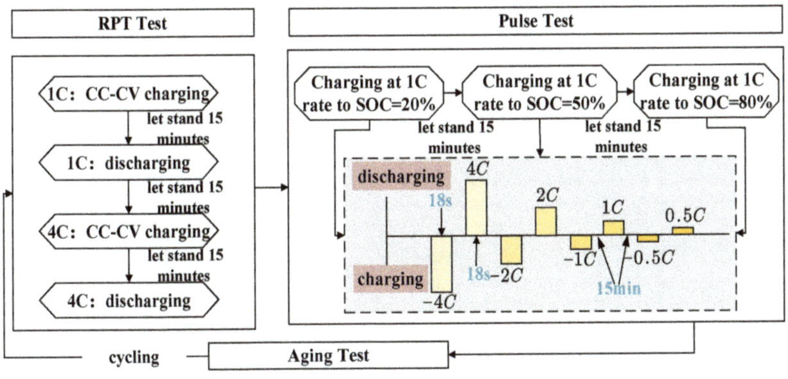

Figure 3. The battery testing process.

The test process is divided into three parts, namely reference performance test (RPT), which is also called capacity test, pulse test and aging test. RPT test is used for capacity calibration. The aging features extracted in this paper come from the battery data in the pulse test stage, while the aging test is used to accelerate the aging of the battery. This paper considers two temperature modes, that is, accelerated aging modes with different temperatures selected during the aging test stage.

Figures 4 and 5 show the current/voltage changes in the lithium-ion battery under three SOC during the pulse current test. It can be seen that the response voltage value changes significantly in different SOC stages, for example, in the charging stage where the SOC changes in the range of 20% to 80%, compared with the two stages of SOC = 20% and SOC = 50%, the response voltage value of the corresponding point is the largest at the stage of SOC = 80%, and the response voltage value of SOC = 50% is larger than that of SOC = 20%. It can be said that when the SOC is in the charging stage belonging to this area, the response voltage value of the corresponding point increases with the increase in the initial SOC value. Under the same conditions, the same results can be obtained in the discharging stage as in the charging stage.

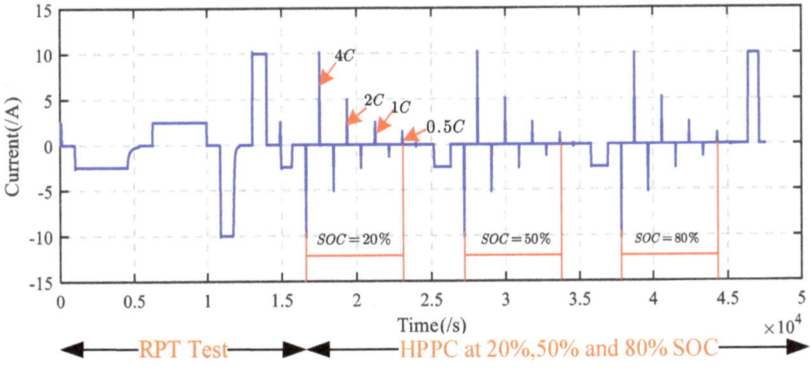

Figure 4. Current change curve under RPT tests and pulse tests of batteries.

Since the pulse current value is always at a stable value, this paper only considers the changing characteristics of the response voltage under the current pulse test. Now, we analyze the specific changing trend of the current pulse test response voltage curve in detail, taking the response voltage curve at the SOC= 20% stage as an example. During the charging stage, Figure 6 shows the lithium-ion battery at SOC= 20% at different rates. From the response voltage change curve, it can be seen that the overall trend of the response voltage curve is downward (showing a "red–green–blue" color change); similarly, the discharge stage under this condition also shows the same trend, as shown in Figure 7.

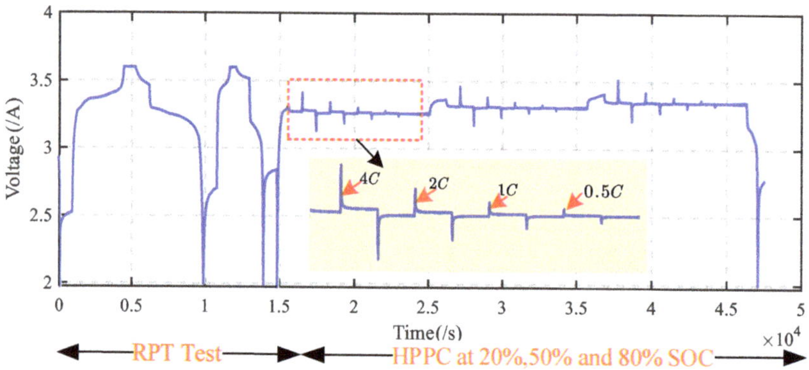

Figure 5. Voltage change curve under RPT tests and pulse tests of batteries.

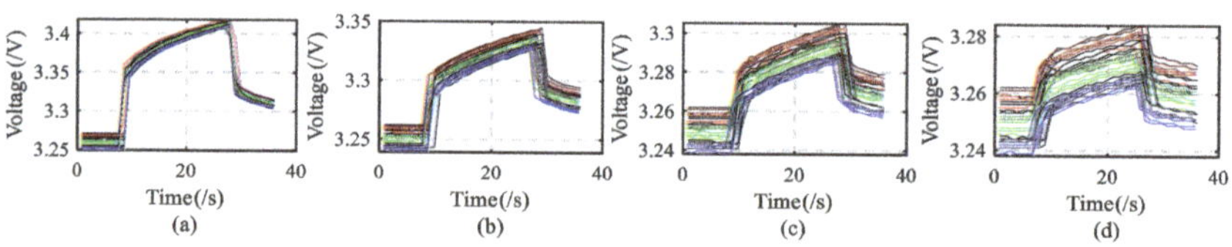

Figure 6. Charging voltage variation curve of the battery under pulse test SOC = 20%. (**a**) 4 C process; (**b**) 2 C process; (**c**) 1 C process; (**d**) 0.5 C process.

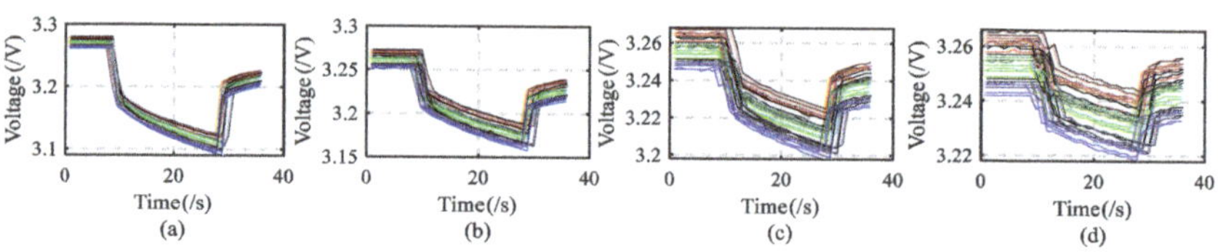

Figure 7. Disharging voltage variation curve of the battery under pulse test SOC = 20%. (**a**) 4 C process; (**b**) 2 C process; (**c**) 1 C process; (**d**) 0.5 C process.

Let us analyze the specific changing trends of the response voltage curves under different initial SOC stages. Taking the charging response voltage curve at a rate of 4 C as an example, Figure 8 shows the response voltage change curve of lithium-ion battery under different initial SOC stages. It can be seen that as the initial value of SOC changes, the voltage response curve changes. When SOC = 20%, the overall trend is downward, but when SOC = 50% and SOC = 80%, the overall trend of the response voltage curve is upward. In the discharging stage under the same conditions, no matter which SOC stage it is in, the overall trend of the response voltage curve is downward, as shown in Figure 9.

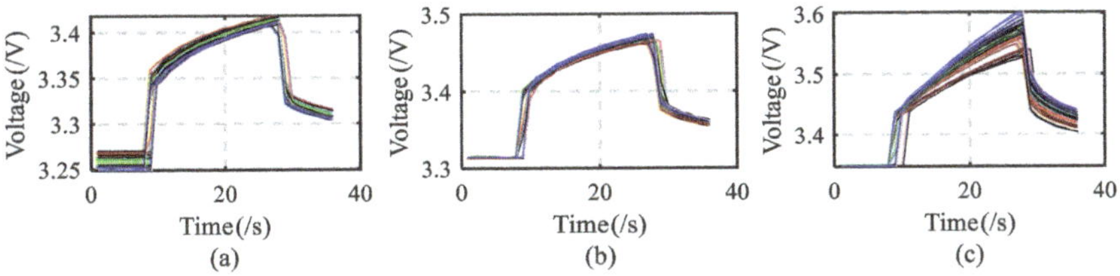

Figure 8. Charging voltage variation curve of the battery under a pulse test rate of 4 C. (**a**) SOC = 20%; (**b**) SOC = 50%; (**c**) SOC = 80%.

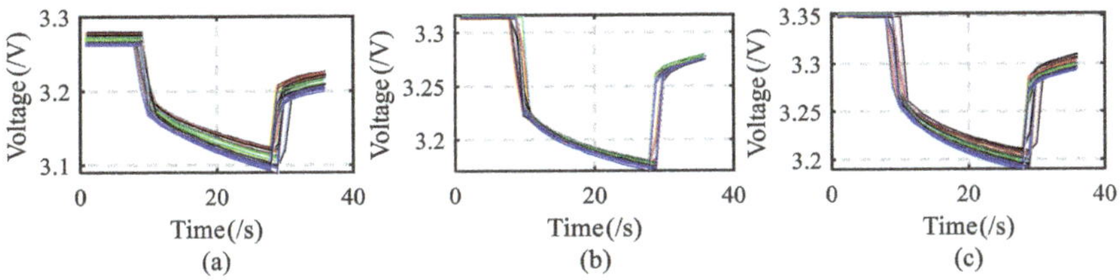

Figure 9. Discharging voltage variation curve of the battery under pulse test rate of 4 C. (**a**) SOC = 20%; (**b**) SOC = 50%; (**c**) SOC = 80%.

The aging features selected in this paper come from the data in the pulse test stage. The purpose of extracting the aging features is to find features that are highly correlated with battery SOH as model input. According to Figures 6–9, the battery voltage shows an obvious downward or upward trend, and it is feasible to perform feature extraction based on this phenomenon.

According to Figure 10, the four voltage values V_1, V_2, V_3, V_4 on the response voltage curve and the integrated area S between the response voltage values V_2 and V_3 are selected as the battery aging features. Based on the above analysis, ten aging features can be obtained at each charging/discharging rate of each SOC stage. For example, during the charging period at a rate of 4 C, there are four response voltage values and one integration area, and there are a total of five aging features. Similarly, there are five aging features during the discharge stage, so there are a total of ten aging features at a rate of 4 C. Therefore, in each SOC stage, there are a total of forty aging features, and by analogy, in the three SOC stages, a total of one hundred and twenty aging features can be obtained.

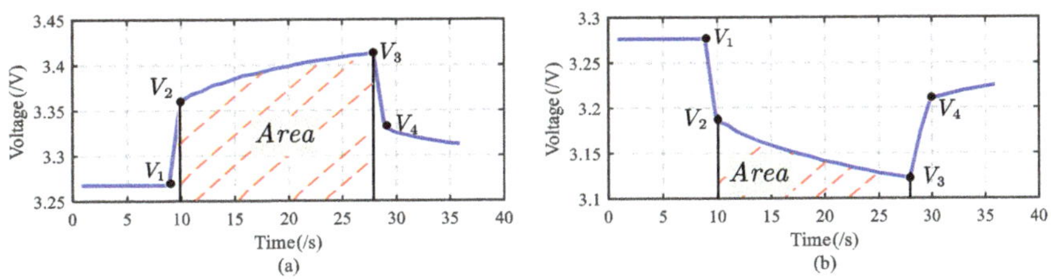

Figure 10. Voltage response curve based on pulse test. (**a**) Charging process; (**b**) discharging process.

3. Lithium-Ion Battery SOH Estimation under Complete SOC Range

Due to the diversity of dynamic working conditions in the battery aging experiment test process, such as differences in charge and discharge rates, variability in operating temperature, and randomness of the current SOC, which increases the types of source domain data and affects the effectiveness of the lithium-ion battery SOH estimation model. On the other hand, the dynamic working conditions of the battery are changing, and the previous battery aging data information is not necessarily applicable to the current new application scenarios. Considering how to select data relevant to the current scenario from diverse source domain data is a prerequisite to ensure accurate estimation of the battery SOH under multiple working conditions.

This paper selects five lithium-ion batteries in two temperature modes as shown in Table 1. First, the battery aging features are extracted from the current pulse test data. Secondly, considering the impact of differences in different battery data on model estimation accuracy, a battery SOH estimation framework based on the sample transfer learning method is established. Then, by constructing an auxiliary data set to enhance the learning ability of the battery SOH model, it can obtain an accurate estimation of the battery SOH under dynamic testing conditions. Finally, comparative experiments are designed and combined with multiple error index methods to verify the effectiveness of the method.

3.1. Lithium-Ion Battery SOH Estimation Framework Based on the Tradaboost.R2 Algorithm

The Tradaboost.R2 algorithm is evolved from the Adaboost.R2 algorithm, which is essentially a machine learning method based on the concept of ensemble learning. The purpose of the Adaboost.R2 algorithm is to update and iterate two weight parameters; one is the weight parameter for the sample data, and the other is the weight parameter for the weak learner. The significance of updating the sample data weight parameters is to adjust the distribution of the training data set for the next iteration based on the previous round for the source domain. In other words, in each iteration process, the weight coefficient for the source domain sample data that makes the estimation result error of the target domain smaller is reduced in the next iteration. And for the source domain sample data that makes the prediction result of the target domain have a larger error, its weight coefficient will be increased in the next iteration, because in the next iteration, these sample data that make the error result larger may make the error result smaller in this iteration, but if the error result does not decrease in this iteration, then it continues to increase the weight of the sample data and proceeds to the next round of iteration. On the other hand, the purpose of updating the weight parameters for the weak learner is to increase the weight coefficient for the weak learner with a smaller overall error result.

Tradaboost.R2 is based on the Adaboost.R2 algorithm, adding the original source domain data and target domain data to the source domain, target domain and auxiliary data set. The auxiliary data set comes from a small amount of labeled data in the target domain. At this time, the training set consists of the source domain data and the auxiliary data. The goal of the Tradaboost.R2 algorithm is to reduce the sample data in the auxiliary data set that makes the prediction result error larger, so that after several iterations, the sample data in the auxiliary set related to the source domain data will have a larger weight coefficient, and the weight coefficient of sample data in the auxiliary data set that is not related to the source domain will be reduced.

This paper chooses extreme learning machines (ELMs) as the weak learner; an ELM is actually a single hidden layer feedforward neural network method. Compared with the traditional feedforward neural network, the structure still includes an input layer, a hidden layer and an output layer. However, in terms of training effect, since the ELM does not require the backpropagation method based on gradient descent to adjust the weight coefficients, its network training speed is fast and it has good generalization ability, while it does not easily fall into local minimum points. Therefore, this paper will use an ELM as the weak learner to establish a lithium-ion battery SOH estimation framework based on the Tradaboost.R2 method.

Consider a large amount of labeled source domain data $D_s = \{(x_{s1}, y_{s1}), \ldots, (x_{sn}, y_{sn})\}$, where $x_{si} \in X_s$ is input data and $y_{si} \in Y_s$ is output data. Dividing the target domain data into two parts, the auxiliary data set X_{t_f} and the testing set X_{t_t}, where the auxiliary data set X_{t_f} is a small amount of labeled data in the target domain, and the corresponding sample label is Y_{t_f}, while the test data set X_{t_t} is a large amount of unlabeled data in the target domain. Now the entire training data consist of the source domain data set and auxiliary data set, that is $X = X_s \cup X_{t_f}$, and the corresponding sample label is $Y = Y_s \cup Y_{t_f}$. Assume that the amount of sample data in the source domain is n_s and the number of samples in the auxiliary data set is n_f. For simplicity of writing, the test data in the target domain are written as X_u. Therefore, the current goal is to predict the output of the corresponding sample through the trained model when the model has a new test sample input $x_i^* \in X_u$. The specific process of the Tradaboost.R2 method is shown in Algorithm 1.

3.2. Battery SOH Estimation Based on Complete SOC Range

Based on the research and analysis of battery aging features in Section 2, it can be seen that 40 battery aging features can be extracted from each SOC interval, and a total of 120 aging features are obtained. In order to establish a battery SOH estimation model under different temperature modes, all the aging features of the three SOC intervals are used as the input of the model, and the application scenarios in single-temperature mode and multi-temperature mode are considered, respectively. Next, considering that the source domain and the target domain come from two different battery data sets, the entire data set is divided into source domain, auxiliary and target domain data sets. By using a small amount of target domain label sample data to establish an auxiliary data set, the impact of insufficient source domain information on the model estimation accuracy is reduced. Table 2 shows the division of source domain, auxiliary and target domain data sets under the complete SOC interval, where T_s represents source domain data, T_f represents auxiliary data and T_t represents target domain data. At the same time, in order to verify the effectiveness of the battery SOH estimation model under different temperature modes, six scenarios are considered and represented by the symbol Si, where $S1$–$S3$ are single-temperature scenarios, that is, the source domain and the target domain have the same temperatures, and $S4$–$S6$ are multi-temperature scenarios, that is, battery data from the source domain and target domain at different temperatures.

Algorithm 1: Tradaboost.R2 algorithm.

Input: training input data set$x_i \in X = X_s \cup X_{t_f}$ $(i = 1, 2, \ldots, n_s, n_s + 1, \ldots, n_s + n_f)$, corresponding label data set $Y = Y_s \cup Y_{t_f}$, unlabeled target domain test data set X_u, weak learner f, number of iterations N, the maximum number of iterations N_{max}.

Output: SOH estimated value for test data $y = \sum_{l=1}^{N_{max}} \beta_l f(X, \theta_l)$.

Parameter initialization: weight coefficient of source domain and auxiliary set sample are, respectively, $w_s^i = \frac{1}{n_s}, w_{t_f}^i = \frac{1}{n_f} (i = 1, 2, \ldots, n_s, n_s + 1, \ldots, n_s + n_f)$; weight coefficient of weak learner $\beta = \frac{1}{1+\sqrt{2ln\frac{n_s}{N}}}$;

for *t=1 to T* **do**

  The training model based on ELM needs to meet the following conditions $min_{\theta_N} \sum_{p=1}^{n_s+n_f} w||y_p - f(x_p, \theta_N)||_2^2$;

  Compute model error $E_p = \frac{(y_p - f(x_p,\theta_N))^2}{max|y_p - f(x_p,\theta_N)|} \quad (p = 1, 2, \ldots, n_s + n_f)$;

  Update sample weights $w_s^i, w_{t_f}^j$ and sub-model weights β_N. **if** $N \leq N_{max}$ **then**

    Update sub-model weights: $\beta_N = \frac{\sigma}{1-\sigma}$, where $\sigma = \begin{cases} \sum_{p=1}^{n_s+n_f} E_p w, & \sum_{p=1}^{n_s+n_f} E_p w < 0.5, \\ 0.5, & \sum_{p=1}^{n_s+n_f} E_p w \geq 0.5. \end{cases}$;

    Update sample weights: $w_s^i = w_s^i \beta_N^{E_i}, w_{t_f}^j = w_{t_f}^j \beta_N^{-E_j}$, where $(i = 1, 2, \ldots, n_s; j = 1, 2, \ldots, n_f; N = N + 1)$

Furthermore, in order to verify the effectiveness of the battery SOH estimated model, the mean absolute error (MAE), mean square error (MSE) and root mean square error (RMSE) analysis methods are used, which are expressed in Equation (1)–(3). Finally, the coefficient of determination R^2 is used to verify the performance of the estimated model, where the larger the R^2 value, the better the model fitting effect.

$$E_{MAE} = \frac{1}{N}\sum_{i=1}^{N}|Y_{i,true} - \bar{Y}_i| \tag{1}$$

$$E_{MSE} = \frac{1}{N}\sum_{i=1}^{N}(Y_{i,true} - \bar{Y}_i)^2 \tag{2}$$

$$E_{RMSE} = \sqrt{\frac{1}{N}\sum_{i=1}^{N}|Y_{i,true} - \bar{Y}_i|} \tag{3}$$

$$R^2 = 1 - \frac{\sum_{i=1}^{N}(Y_{i,true} - \bar{Y}_i)^2}{\sum_{i=1}^{N}(Y_{i,true} - Y_{mean})^2} \tag{4}$$

where N is the number of samples, and $Y_{i,true}$ is the true value for battery SOH, $\bar{Y}_i$ is the estimated value for battery SOH, and the Y_{mean} is the mean value for battery SOH.

Table 2. Division of source domain, auxiliary and target domain data sets under complete SOC intervals.

Data	Single Temperature			Multiple Temperature		
	S1	S2	S3	S4	S5	S6
T_s	No.1	No.2	No.5	No.1	No.2	No.3
T_f	No.3 (15%)	No.3 (15%)	No.4 (30%)	No.4 (15%)	No.4 (15%)	No.4 (15%)
T_t	No.3 (85%)	No.3 (85%)	No.4 (70%)	No.4 (85%)	No.4 (85%)	No.4 (85%)

This paper selects the Adaboost method, GPR, SVM and CNN-LSTM, as well as the case where the auxiliary set is not considered in the proposed method as the comparison methods. The symbol *Adaboost* represents the estimated results based on the Adaboost.R2 method, the symbol *GPR* represents the estimated results based on the GPR method, the symbol *SVM* represents the estimated results based on the SVM method, the symbol *CNN-LSTM* represents the estimated results based on the CNN-LSTM method, the symbol *No-Transfer* represents the estimated results without the transfer learning step (that is, the proposed method does not include the auxiliary data), and the symbol *Transfer* represent the estimated results of the method which proposed in this paper.

Due to the different temperatures of lithium-ion batteries, the entire battery data can be divided into two modes, namely mode 1 and mode 2. Regarding the division of data sets in single-temperature mode, take the three battery data sets in mode 1 as an example. There are three battery data sets in mode 1, namely No.1, No.2 and No.3. Choose one of these three data sets as the target domain, and the other two data sets as the source domain, such as scenario 1 (*S*1) and scenario 2 (*S*2). In the multi-temperature mode, select the data set in one mode as the source domain, such as No.1 in mode 1, and the data set in another temperature mode as the target domain, such as No.4 in mode 2, and divide it into scenarios 4, 5 and 6 (*S*4, *S*5, *S*6). In these two modes, in order to improve the estimation accuracy of the model and verify the necessity of transfer learning, the construction of auxiliary data sets is considered, as shown in Table 2.

3.3. Analysis and Discussion of Experimental Results

In this part, based on the previous aging feature extraction and model optimization, the battery SOH estimation under single-temperature and multi-temperature modes are analyzed. First, the estimation results under single-temperature mode are analyzed, and the experimental results based on different models are considered to verify the necessity of transfer learning.

Figures 11 and 12 show the battery SOH estimation results for scenario 1 (*S*1)–scenario 3 (*S*3) under the single-temperature mode. It can be found that in the three scenarios, the battery SOH estimation based on the *Transfer* method are closer to the reference value, and its effect of tracking the reference value is more obvious, while the battery SOH estimation based on the*No-Transfer* method obviously does not accurately track the true value. In addition, considering that the Tradaboost.R2 algorithm is improved from the Adaboost.R2 algorithm, the battery SOH estimation based on the basic algorithm Adaboost is also unsatisfactory. At the same time, regardless of the scenario, the trend of estimation results based on CNN-LSTM, GPR and SVM is not ideal.

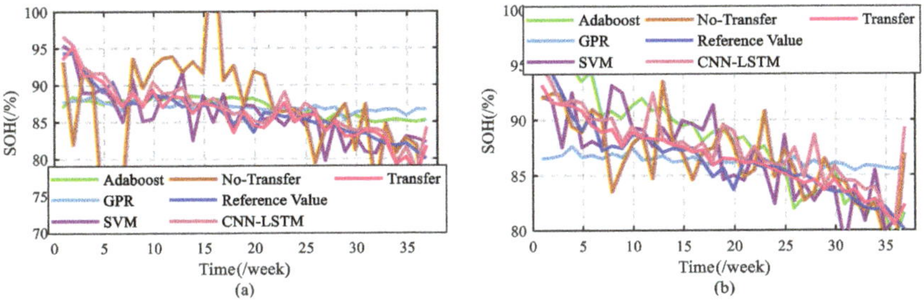

Figure 11. Battery SOH estimation in single-temperature mode. (**a**) Scenario 1; (**b**) scenario 2.

Tables 3 and 4 give three error values and coefficients of determination for *S*1–*S*3. It can be seen from these two tables that the MAE of the estimated value based on the *Transfer* method does not exceed 1.5%, which is significantly smaller than the MAE of other comparison methods; in terms of MSE, the MSE based on the *Transfer* method is lower than 0.015% and the corresponding RMSE is less than 1.5%, which also shows that the estimation accuracy based on the *Transfer* method is higher than the other methods. On the other hand, in terms of verifying the model fitting effect, the coefficient of determination R^2 based on the *Transfer* method is not less than 88% in the three scenarios. Compared with the corresponding coefficient of determination values of other methods, the model fitting effect based on the *Transfer* method is better. Therefore, it can be seen that the method after the transfer learning step can improve the accuracy of the estimation results, but the estimation results without the transfer learning step cannot guarantee the estimation accuracy, which further illustrates the necessity of transfer learning.

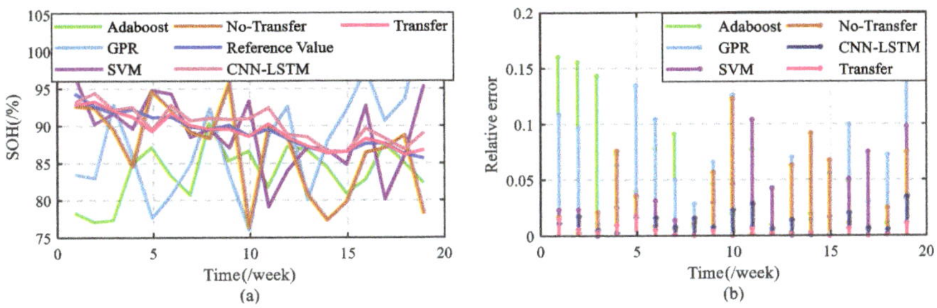

Figure 12. Battery SOH estimation in single-temperature mode under scenario 3. (**a**) SOH estimation; (**b**) relative error.

On the other hand, according to Table 2, it can be seen that when no auxiliary set is constructed, taking *S*1 as an example, the MAE is 0.0503, which shows that the estimation model obtained only with the help of source domain data information is not well adapted to the test environment of the target domain. When constructing the auxiliary data set, only the first 15% of the label data information in the target domain can obtain better estimation results to verify the necessity of transfer learning. It can be seen that lithium-ion battery SOH estimation can be completed with a small amount of label sample data. Therefore, this method verifies that the accuracy of battery SOH estimation based on a small number of label sample data is not affected.

Table 3. Three error values and coefficients of determination for scenario 1 and scenario 2.

Model	S1				S2			
	MAE	MSE	RMSE	R^2	MAE	MSE	RMSE	R^2
Adaboost	0.0211	7.48×10^{-4}	0.0273	0.2178	0.0223	6.59×10^{-4}	0.0257	0.7780
GPR	0.0229	9×10^{-4}	0.03	0.1063	0.0222	8.83×10^{-4}	0.0297	0.0268
SVM	0.0186	5.19×10^{-4}	0.0228	0.7209	0.0212	7.02×10^{-4}	0.0265	0.6753
No-Transfer	0.0503	0.0051	0.0716	0.522	0.0199	6.89×10^{-4}	0.0262	0.6497
CNN-LSTM	0.0158	2.36×10^{-4}	0.0154	0.8178	0.0118	3.41×10^{-4}	0.0185	0.7523
Transfer	0.0103	1.5×10^{-4}	0.0123	0.8965	0.0079	1.12×10^{-4}	0.0106	0.8896

Table 4. Three error values and coefficients of determination for scenario 3 and scenario 4.

Model	S3				S4			
	MAE	MSE	RMSE	R^2	MAE	MSE	RMSE	R^2
Adaboost	0.0568	0.0057	0.0754	0.4371	0.0342	0.0018	0.0428	0.2637
GPR	0.0722	0.0073	0.0855	0.4084	0.1045	0.0173	0.1314	0.5953
SVM	0.0329	0.0020	0.0448	0.5383	0.0327	0.0017	0.0410	0.3319
No-Transfer	0.0371	0.0026	0.0512	0.5959	0.0553	0.0050	0.0706	0.5073
CNN-LSTM	0.01	1.52×10^{-4}	0.0123	0.7887	0.0257	0.0011	0.0325	0.7310
Transfer	0.0051	5.30×10^{-5}	0.0073	0.8981	0.0108	2.32×10^{-4}	0.0152	0.9045

Besides that, Tables 3 and 4 present the error estimation results of the battery SOH under different models. These tables also depict the error estimation results of the battery SOH under different types. In order to analyze the estimated result clearly, we use the estimated results based on GPR, SVM, CNN-LSTM, Adaboost.R2 and without using the transfer method under the same conditions to compare with the Tradaboost.R2 algorithm estimated results.

After analyzing the battery SOH estimation results in single-temperature mode, this part starts to analyze the multi-temperature mode. Battery No.5 in mode 2 only has ten sample points that select battery No.4 in mode 2 as the target domain, and the three batteries No.1, No.2 and No.3 in mode 1 are used as the source domain to verify the battery SOH estimation under multi-temperature mode.

Figures 13 and 14 show the battery SOH estimation results for scenario 4 (*S*4)–scenario 6 (*S*6) under multiple-temperature mode. It can be found that in the three scenarios, the battery SOH estimation based on the *Transfer* method is closer to the reference value, and its effect of tracking the reference value is more obvious, while the battery SOH estimation based on the *No-Transfer* method obviously does not accurately track the true value. In addition, considering that the Tradaboost.R2 algorithm is improved from the Adaboost.R2 algorithm, the battery SOH estimation based on the basic algorithm Adaboost is also unsatisfactory. At the same time, regardless of the scenario, the trend of estimation results based on CNN-LSTM, GPR and SVM is not ideal.

Tables 4 and 5 give three error values and coefficients of determination for scenario *S*4–*S*6. It can be seen from these two tables that the MAE of the battery SOH estimated value based on the *Transfer* method does not exceed 1.3%, which is significantly smaller than the MAE of other comparison methods; in terms of MSE, the MSE based on the *Transfer* method is lower than 0.033% and the corresponding RMSE is less than 1.8%, which also shows that the estimation accuracy based on the *Transfer* method is higher than that of other methods. On the other hand, in terms of verifying the model fitting effect, the coefficient of determination R^2 based on the *Transfer* method is not less than 85% in the three scenarios.

Compared with the corresponding coefficient of determination values of other methods, the model fitting effect based on the *Transfer* method is better than other methods.

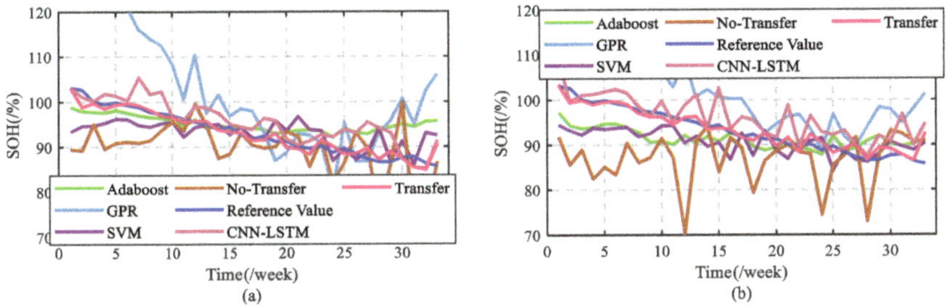

Figure 13. Battery SOH estimation in multiple-temperature mode. (**a**) Scenario 4; (**b**) scenario 5.

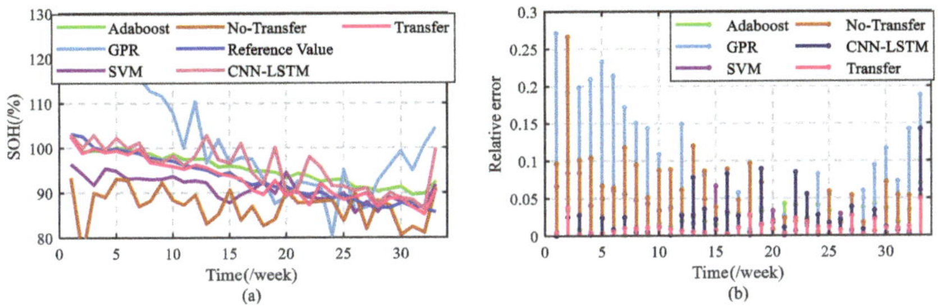

Figure 14. Battery SOH estimation in multiple-temperature mode under scenario 6. (**a**) SOH estimation; (**b**) relative error.

In the three scenarios under multi-temperature mode, taking scenario *S*4 as an example, when the auxiliary data set is not constructed, the MAE is 0.0553, compared with the MAE value (0.0108) under the method proposed in this paper, which shows that the estimation model obtained only with the help of source domain information has no way to learn the data information for the target domain, and when the top 15% of the label data in the target domain is used to build an auxiliary data set, the error value of the prediction results has dropped significantly. This shows that 15% of labeled data information plays an important role in model training and verifies the necessity of transferring data information. The same results are also reflected in scenarios *S*5 and *S*6, which means that when the auxiliary data set only accounts for 15% of the target domain, the accuracy of battery SOH estimation under multi-temperature mode still can be guaranteed.

Table 5. Three error values and coefficients of determination for scenario 5 and scenario 6.

Model	S5				S6			
	MAE	MSE	RMSE	R^2	MAE	MSE	RMSE	R^2
Adaboost	0.0356	0.0017	0.0417	0.3034	0.0240	8.11×10^{-4}	0.0285	0.6668
GPR	0.0908	0.0106	0.1028	0.5998	0.0992	0.0156	0.1249	0.6019
SVM	0.0327	0.0017	0.0410	0.3319	0.0313	0.0016	0.0397	0.4618
No-Transfer	0.0553	0.0050	0.0706	0.5073	0.0666	0.0068	0.0822	0.4413
CNN-LSTM	0.0287	0.0013	0.0364	0.6725	0.0267	0.0015	0.0381	0.6589
Transfer	0.0123	3.21×10^{-4}	0.0179	0.8583	0.0125	2.66×10^{-4}	0.0163	0.8872

4. Lithium-Ion Battery SOH Estimation under Different SOC Ranges

In the previous section, we discuss the battery SOH estimation problem under the complete SOC interval, which means that when establishing the estimation model framework, all aging features are used as input of the model. However, if there is a lack of aging features in a certain range, will it make it difficult to estimate the batter SOH? Under laboratory conditions, pulse test data for three complete SOC intervals can be easily obtained. However, in actual applications, depending on the needs of different application scenarios, it may not be possible to complete the pulse test under three complete SOC intervals. At this time, it is necessary to consider whether the battery SOH estimation under different SOC intervals can be achieved. In other words, comparing with the battery SOH estimation under the complete SOC interval, can it be ensured that the mapping relationship under different SOC intervals will not be affected.

4.1. Data Set Division under Different SOC Intervals

In order to illustrate the aging features under different SOC intervals, the symbol I_1 is used to represent the forty aging features at SOC=20%, the symbol I_2 is used to represent the forty aging features at SOC=50% and the symbol I_3 is used to represent the forty aging features at SOC=80%.

In order to verify that the mapping relationship of the battery SOH estimation model under different SOC intervals is not affected, aging feature data from different initial SOC is selected when constructing the auxiliary and the source domain data set. For example, it is now necessary to use the label data information of battery No.1 to estimate SOH of battery No.3. If battery No.3 only contains pulse test data with SOC=20%, and battery No.1 contains pulse test data with a complete SOC range, then the target domain data set of the model is the aging features I_1 of battery No.3, and source domain data set contains the aging characteristics $I_1 \cup I_2 \cup I_3$ of battery No.1. Under the assumption of this scenario condition, the dimensions of the source domain and target domain data of the model are different. Therefore, the aging feature I_1 of battery No.1 can be selected as the auxiliary data set of the model to learn the data information of I_1, and the aging feature I_2 or I_3 can be used as the source domain data set of the model, thus ensuring the dimensions of the three data sets are the same, and the selection of the auxiliary data set only requires aging features from the same SOC interval as the target domain data set. At the same time, there is no restriction that the auxiliary data must come from the same dataset as the target domain. Therefore, based on the battery pulse test data in different SOC intervals, in each scenario, the source domain, auxiliary data and target domain data can be divided as shown in Table 6.

Table 6. Division of source domain, auxiliary and target domain data sets under different SOC intervals.

	*M*1	*M*2	*M*3	*M*4	*M*5	*M*6
T_s	I_2	I_3	I_1	I_3	I_2	I_1
T_f	I_1	I_1	I_2	I_2	I_3	I_3
T_t	I_1	I_1	I_2	I_2	I_3	I_3

4.2. Battery SOH Estimation Based on Different SOC Ranges

When considering the battery SOH estimation based on sample transfer learning under different SOC intervals, six scenarios are considered and are divided as shown in Table 7. Scenario 1 ($S1$)–Scenario 3 ($S3$) verify the battery SOH estimation based on a single-temperature mode, that is, although the source domain and the target domain data set come from different battery data, they remain in the same temperature mode. The battery SOH estimation under multi-temperature mode is reflected by Scenario 4 ($S4$)–Scenario 6 ($S6$), that is, the source domain and the target domain data set are derived from different temperature modes. The purpose is to verify the performance based on the idea that when using data in different temperature modes, the battery SOH estimation can still be achieved. In each scenario of this part, the source domain, auxiliary data and target domain data set are divided in six ways. Different from the construction method of the auxiliary data set under the complete SOC interval, here, there is no requirement that the auxiliary data must come from the same target domain data set. It is only required that the auxiliary data set and the target domain data set come from the aging features under the same SOC interval. This greatly reduces the demand for target domain label data and has more practical research significance.

Table 7. Division of source domain and target domain data sets under different SOC intervals.

Data	Single Temperature			Multiple Temperature		
	S1	S2	S3	S4	S5	S6
T_s	No.1	No.2	No.5	No.1	No.2	No.3
T_t	No.3	No.3	No.4	No.4	No.4	No.4

4.3. Analysis and Discussion of Experimental Results

In this section, the battery SOH estimation under single-temperature and multi-temperature modes are analyzed, respectively. First, the experimental results under the single-temperature mode are analyzed. Based on Section 3.2, the battery SOH estimation under the six division methods is verified in each scenario; secondly, compared with the battery SOH estimation framework under the complete SOC interval, the feature dimension of the model is not only reduced but also the demand for target domain label data is reduced. Therefore, in this section, the estimation results under the six partitioning methods will be compared with the estimation results based on the *Transfer* method in the complete SOC interval of Section 3.2. In addition, since the Tradaboost.R2 algorithm is an ensemble method, considering the final estimation result is synthesized based on other sub-models, the relative error between each sub-model and the final result is analyzed. Similarly, the battery SOH estimation under multi-temperature mode is also analyzed in the same way.

Figures 15 and 16 show the battery SOH estimation results for scenario 1 ($S1$)–scenario 3 ($S3$). It can be seen from Figure 15 that in $S1$ and $S2$, compared with the estimation results for the other five methods, the estimation results based on method 6 ($M6$) are closer to the reference value, and its effect of tracking the reference value is more obvious. The absolute error also can reflect the better estimation effect is shown in Figure 16 (b).

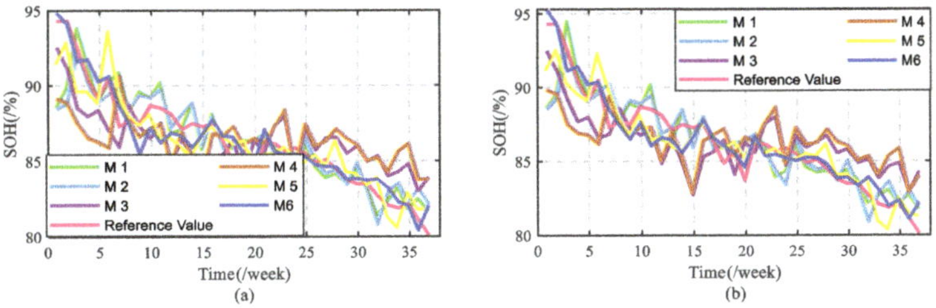

Figure 15. Battery SOH estimation in single-temperature mode. (**a**) Scenario 1; (**b**) scenario 2.

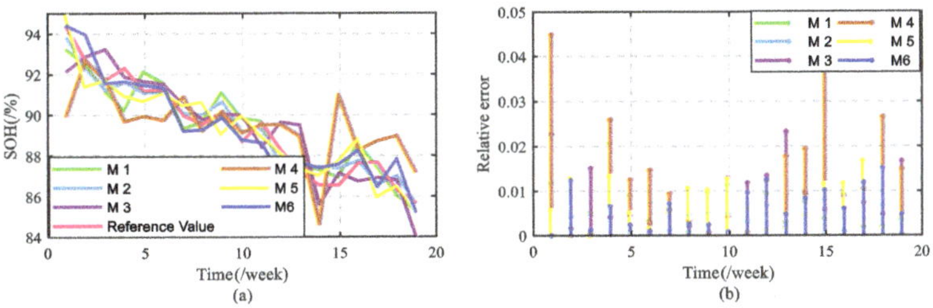

Figure 16. Battery SOH estimation in single-temperature mode under scenario 3. (**a**) SOH estimation; (**b**) relative error.

Tables 8 and 9 give three error values and coefficients of determination for $S1$–$S3$. It can be seen from these two tables that the optimal estimation results can be obtained based on $M6$ in $S1$ and $S2$, while the optimal results in $S3$ are obtained by $M2$. In the optimal estimation results of these three scenarios, the MAE does not exceed 0.75%, which is significantly smaller than the MAE under the other methods; at the same time, the MSE is lower than 0.0095% and the corresponding RMSE does not exceed 0.98%; compared with the complete SOC interval based on $Transfer$ method, the optimal estimation errors of the three scenarios are smaller than the estimation errors based on the $Transfer$ method. On the other hand, in terms of verifying the model fitting effect, the coefficient of determination R^2 of the optimal result reaches 90% in all three scenarios. Compared with the coefficient of determination value based on the $Transfer$ method, it can be seen that the model fitting effect based on different SOC intervals is better than the model effect based on the complete SOC interval. This can also show that while the feature dimension and the demand for label data in the target domain are reduced, the battery SOH estimation accuracy has not been reduced.

Tables 8 and 9 present the error estimation results of the battery SOH under different SOC ranges. In order to ensure that the mapping relationship of the lithium-ion battery SOH estimation model in different SOC intervals is not affected, it adopts the way of dividing the data set in Tables 6 and 7.

After analyzing the battery SOH estimation results in single-temperature mode, this part starts to analyze the multi-temperature mode. We select battery No.4 in mode 2 as the target domain and use the three batteries No.1, No.2 and No.3 in mode 1 as the source domain to verify the battery SOH estimation under multi-temperature mode.

Figures 17 and 18 show the battery SOH estimation results for scenario 4 ($S4$)–scenario 6 ($S6$). It can be seen from Figure 17 that in $S4$ and $S5$, comparing with the estimation results of the other five methods, the estimation results based on method 5 ($M5$) are closer to the reference value, and its effect of tracking the reference value is more obvious, which the absolute error also can reflect the better estimation effect that shown in Figure 18 (b). Completing the verification of the estimation results of three scenarios under the multi-temperature mode, which demonstrates that the method proposed in this paper has stronger generalization ability in estimating battery SOH under different temperature modes.

Table 8. Three error values and coefficients of determination for scenario 1 and scenario 2.

Model	S1				S2			
	MAE	MSE	RMSE	R^2	MAE	MSE	RMSE	R^2
$M1$	0.0123	2.96×10^{-4}	0.0172	0.7576	0.0113	2.74×10^{-4}	0.0166	0.7524
$M2$	0.0130	3.32×10^{-4}	0.0182	0.6934	0.0119	3.04×10^{-4}	0.0174	0.7089
$M3$	0.0198	4.92×10^{-4}	0.0222	0.4473	0.0198	4.97×10^{-4}	0.0223	0.4396
$M4$	0.0238	7.78×10^{-4}	0.0279	0.2215	0.0226	6.98×10^{-4}	0.0264	0.2618
$M5$	0.0107	1.77×10^{-4}	0.0133	0.8269	0.0096	1.51×10^{-4}	0.0123	0.8475
$M6$	0.0075	9.48×10^{-5}	0.0097	0.9143	0.0072	8.78×10^{-5}	0.0094	0.9183
$Transfer$	0.0103	1.5×10^{-4}	0.0123	0.8965	0.0079	1.12×10^{-4}	0.0106	0.8896

Table 9. Three error values and coefficients of determination for scenario 3 and scenario 4.

Model	S3				S4			
	MAE	MSE	RMSE	R^2	MAE	MSE	RMSE	R^2
$M1$	0.0065	6.66×10^{-5}	0.0082	0.8811	0.0174	4.73×10^{-4}	0.0217	0.7290
$M2$	0.0047	3.04×10^{-5}	0.0055	0.9397	0.0188	5.59×10^{-4}	0.0237	0.7009
$M3$	0.0092	1.29×10^{-4}	0.0114	0.8324	0.0277	0.0013	0.0364	0.4869
$M4$	0.0142	3.74×10^{-4}	0.0193	0.4273	0.0326	0.0016	0.0406	0.2945
$M5$	0.0073	7.81×10^{-5}	0.0088	0.8737	0.0094	2.00×10^{-4}	0.0142	0.9094
$M6$	0.0064	6.18×10^{-5}	0.0079	0.9043	0.0102	1.52×10^{-4}	0.0124	0.9406
$Transfer$	0.0051	5.30×10^{-5}	0.0073	0.8981	0.0108	2.32×10^{-4}	0.0152	0.9045

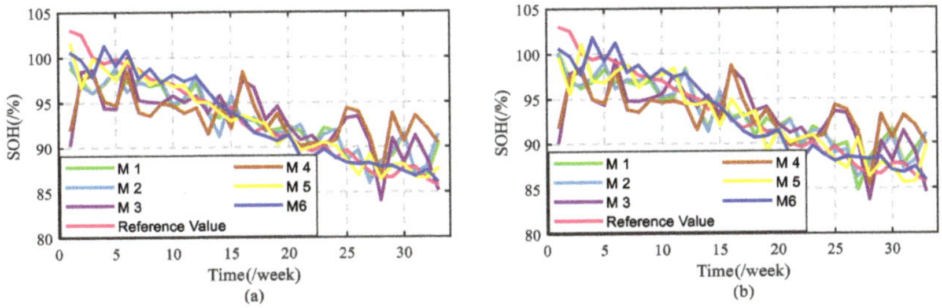

Figure 17. Battery SOH estimation in multiple-temperature mode. (**a**) Scenario 4; (**b**) scenario 5.

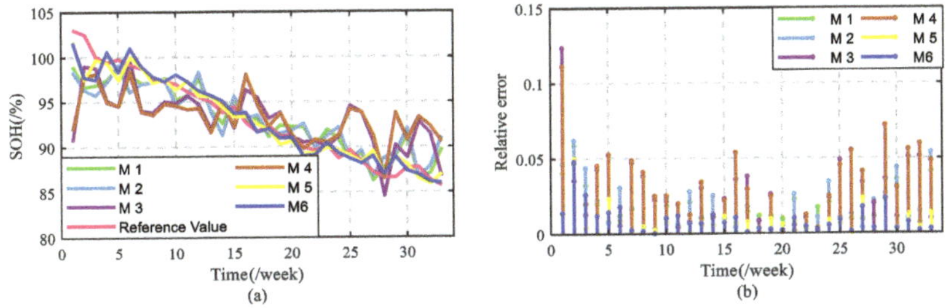

Figure 18. Battery SOH estimation in multiple-temperature mode under scenario 6. (**a**) SOH estimation; (**b**) relative error.

Tables 9 and 10 give three error values and coefficients of determination for *S*4–*S*6. It can be seen from these two tables that the optimal estimation results can be obtained based on *M*6 in *S*4 and *S*5, while the optimal results in *S*6 are obtained by *M*5. In the optimal estimation results of these three scenarios, the MAE does not exceed 1.2%, which is significantly smaller than the MAE under other methods; at the same time, the MSE is lower than 0.02% and the corresponding RMSE does not exceed 1.4%. Compared with the complete SOC interval based on the *Transfer* method, the optimal estimation errors of the three scenarios are smaller than the estimation errors based on the *Transfer* method. On the other hand, in terms of verifying the model fitting effect, the coefficient of determination R^2 of the optimal result reaches 92% in all three scenarios. Therefore, no matter in terms of battery SOH estimation accuracy or model fitting effect, the optimal results based on different SOC intervals are better than complete SOC intervals.

Table 10. Three error values and coefficients of determination for scenario 5 and scenario 6.

Model	*S*5				*S*6			
	MAE	MSE	RMSE	R^2	MAE	MSE	RMSE	R^2
*M*1	0.0195	5.66×10^{-4}	0.0238	0.6905	0.0165	4.27×10^{-4}	0.0207	0.7704
*M*2	0.0171	4.89×10^{-4}	0.0221	0.7365	0.0199	6.40×10^{-4}	0.0253	0.6576
*M*3	0.0285	0.0014	0.0374	0.4889	0.0287	0.0014	0.0371	0.4326
*M*4	0.0327	0.0016	0.0403	0.3162	0.0328	0.0016	0.0406	0.2835
*M*5	0.0157	4.38×10^{-4}	0.0209	0.8155	0.0090	1.69×10^{-4}	0.0130	0.9249
*M*6	0.0114	1.90×10^{-4}	0.0138	0.9294	0.0093	1.73×10^{-4}	0.0132	0.9290
Transfer	0.0123	3.21×10^{-4}	0.0179	0.8583	0.0125	2.66×10^{-4}	0.0163	0.8872

Based on the above discussion, it can be proved from three aspects that the method proposed in this paper can achieve higher accuracy in battery SOH estimation:

(1) In the comparison model that combines various errors to depict the performance of the method, we use the estimated results based on GPR, SVM, CNN-LSTM, Adaboost and without using the transfer method under the same conditions to

compare with the Tradaboost.R2 algorithm estimated results. The various error results show that the estimated results based on the Tradaboost.R2 algorithm are better than other algorithms.

(2) The performance of model fitting is measured by the coefficient of determination R^2. The better the model fitting performance, the closer its value is to 1. The values of the coefficient of determination R^2 prove that the estimated results based on the Tradaboost.R2 algorithm are better than other algorithms.

(3) This paper proposes a lithium-ion battery SOH estimation framework based on the Tradaboost.R2 method under pulse testing. The proposed method can ensure that the mapping relationship of the lithium-ion battery SOH estimation model in different SOC intervals is not affected.

(4) By verifying the estimation results based on the complete SOC interval and different SOC intervals, it can be seen that while the model feature dimension and the demand for target domain label data is reduced, the estimation accuracy of the model can still be guaranteed.

5. Conclusions and Future work

In order to estimate the lithium-ion battery SOH under different dynamic working conditions, this paper establishes a battery SOH estimation model based on the Tradaboost.R2 algorithm. Considering the shortened time to obtain aging features under dynamic test conditions, the aging features that can reflect the attenuation of battery capacity are extracted from the 18s pulse current. At the same time, five lithium-ion batteries are tested in two temperature modes. For the cyclic aging test, the weight of the source domain data samples is adjusted through the Tradaboost.R2 algorithm. Under laboratory conditions, two verification scenarios for lithium-ion battery SOH estimation are designed, which verified that while the age feature dimension is reduced and the demand for target domain label data is reduced, the mapping relationship of the lithium-ion battery SOH estimation model is not affected by the initial value of the SOC.

Considering the impact of aging features on model estimation results, in future work, a framework for automatically extracting battery aging features under dynamic conditions will be established while ensuring that the extracted aging features have a strong correlation with battery SOH. In addition, adaptive online learning should be considered in the future such that the battery model can be adjusted adaptively to achieve the best state whenever new data are available.

Author Contributions: Conceptualization, Y.L. and X.H.; methodology, Y.L. and J.M.; validation, Y.L. and X.H.; software, Y.L. and K.S.; validation, Y.L. and X.H.; supervision, K.S., R.T. and D.I.S.; formal analysis, Y.L.; writing—review and editing, Y.L. and X.H.; writing—original draft preparation, Y.L. and D.I.S. All authors have read and agreed to the published version of the manuscript.

Funding: This work was supported by "the Fundamental Research Funds for the Central Universities", Southwest Minzu University (Grant ZYN2023074), National Natural Science Foundation of China under Grant 52107229, Shaanxi Province Qinchuangyuan High-Level Innovation and Entrepreneurship Talent Project under Grant QCYRCXM-2023-112.

Data Availability Statement: The data are not publicly available due to privacy constraints.

Conflicts of Interest: The authors declare no conflicts of interest.

References

1. Tran, D.D.; Vafaeipour, M.; Baghdadi, M.E.; Barrero, R.; Mierlo, J.V.; Hegazy, O. Thorough state of the mart analysis of electric and hybrid vehicle powertrains: Topologies and integrated energy management strategies. *Renew. Sustain. Energy Rev.* **2020**, *119*, 109596. [CrossRef]
2. Swierczynski, M.J.; Stroe, D.I.; Rasmus, L.; Stan, A.I.; Kjær, P.; Teodorescu, R.; Kær, S.K. Field experience from li-ion bess delivering primary frequency regulation in the danish energy market. *ECS Trans.* **2014**, *61*, 1–14. [CrossRef]
3. Xiong, R.; Zhang, Y.Z.; Wang, J.; He, H.W.; Peng, S.M.; Pecht, M. Lithium-ion battery health prognosis based on a real battery management system used in electric vehicles. *IEEE Trans. Veh. Technol.* **2019**, *68*, 4110–4121. [CrossRef]

4. Zhang, C.L.; Luo, L.J.; Yang, Z.; Zhao, S.S.; He, Y.G.; Wang, X.; Wang, H.X. Battery SOH estimation method based on gradual decreasing current, double correlation analysis and GRU. *Green Energy Intell. Transp.* **2023**, *2*, 100108. [CrossRef]
5. Zhang, Y.L.; Li, X.; Li, Z.H.; Yang, F.Q. Evaluation of electrochemical performance of supercapacitors from equivalent circuits through cyclic voltammetry and galvanostatic charge/discharge. *J. Energy Storage* **2024**, *86*, 111122. [CrossRef]
6. Verma, P.; Maire, P.; Novák, P. A review of the features and analyses of the solid electrolyte interphase in li-ion batteries. *Electrochim. Acta* **2010**, *55*, 6332–6341. [CrossRef]
7. Laayouj, N.; Hicham, J. Lithium-ion battery degradation assessment and remaining useful life estimation in hybrid electric vehicle. *Renew. Energy Sustain. Dev.* **2016**, *2*, 37–44. [CrossRef]
8. Aderyani, S.; Flouda, P.; Shah, S.A.; Green, M.J.; Lutkenhaus, J.L.; Ardebili, H. Simulation of cyclic voltammetry in structural supercapacitors with pseudocapacitance behavior. *Electrochim. Acta* **2021**, *390*, 138822. [CrossRef]
9. Doyle, M.; Fuller, T.F.; Newman, J. Modeling of galvanostatic charge and discharge of the lithium/polymer/insertion cell. *J. Electrochem. Soc.* **1993**, *140*, 1526–1533. [CrossRef]
10. Jiang, B.; Dai, H.F.; Wei, X.Z.; Xu, T.J. Joint estimation of lithium-ion battery state of charge and capacity within an adaptive variable multi-timescale framework considering current measurement offset. *Appl. Energy* **2019**, *253*, 113619. [CrossRef]
11. Liu, K.L.; Shang, Y.; Ouyang, Q.; Widanage, W.D. A data-driven approach with uncertainty quantification for predicting future capacities and remaining useful life of lithium-ion battery. *IEEE Trans. Ind. Electron.* **2021**, *68*, 3170–3180. [CrossRef]
12. Chen, X.W.; Liu, Z.; Sheng, H.M.; Wu, K.P.; Mi, J.H.; Li, Q. Transfer learning based remaining useful life prediction of lithium-ion battery considering capacity regeneration phenomenon. *J. Energy Storage* **2024**, *78*, 109798. [CrossRef]
13. Chen, J.X.; Hu, Y.J.; Zhu, Q.; Rashid, H.; Li, H.K. A novel battery health indicator and PSO-LSSVR for LiFePO4 battery SOH estimation during constant current charging. *Energy* **2023**, *282*, 128782. [CrossRef]
14. Zhang, C.L.; Luo, L.J.; Yang, Z.; Du, B.L.; Zhou, Z.H.; Wu, J.; Chen, L.P. Flexible method for estimating the state of health of lithium-ion batteries using partial charging segments. *Energy* **2024**, *295*, 131009. [CrossRef]
15. Sun, J.H.; Kainz, J. State of health estimation for lithium-ion batteries based on current interrupt method and genetic algorithm optimized back propagation neural network. *J. Power Sources* **2024**, *591*, 233842. [CrossRef]
16. Ko, C.J.; Chen, K.C. Using tens of seconds of relaxation voltage to estimate open circuit voltage and state of health of lithium ion batteries. *Appl. Energy* **2024**, *357*, 122488. [CrossRef]
17. Chen, X.W.; Liu, Z. A long short-term memory neural network based Wiener process model for remaining useful life prediction. *Reliab. Eng. Syst. Saf.* **2022**, *226*, 108651. [CrossRef]
18. Hu, X.; Jiang, H.; Feng, F.; Liu, B. An enhanced multi-state estimation hierarchy for advanced lithium-ion battery management. *Appl. Energy* **2020**, *257*, 114019. [CrossRef]
19. Qian, C.; Xu, B.H.; Xia, Q.; Ren, Y.; Sun, B.; Wang, Z.L. SOH prediction for Lithium-Ion batteries by using historical state and future load information with an AM-seq2seq model. *Appl. Energy* **2023**, *336*, 120793. [CrossRef]
20. Zhang, C.L.; Zhao, S.S.; Yang, Z.; Chen, Y. A reliable data-driven state-of-health estimation model for lithium-ion batteries in electric vehicles. *Front. Energy Res.* **2022**, *10*, 1013800. [CrossRef]
21. Sun, Y.H.; Jou, H.L.; Wu, J.C. Auxiliary diagnosis method for lead-acid battery health based on sample entropy. *Energy Convers. Manag.* **2009**, *50*, 2250–2256. [CrossRef]
22. Li, Y.Y.; Stroe, D.I.; Cheng, Y.H.; Sheng, H.M.; Sui, X.; Teodorescu, R. On the feature selection for battery state of health estimation based on charging-discharging profiles. *J. Energy Storage* **2021**, *33*, 102122. [CrossRef]
23. He, Y.J.; Chen, Y.A.; Chen, N.; Xie, Y.H.; Wang, H.; Huang, W.; Zhao, X.; Yang, C.H. State of health estimation of lithium-ion battery aging process based on time-frequency fusion characteristics. *J. Power Sources* **2024**, *596*, 234002. [CrossRef]
24. Qian, C.; Guan, H.S.; Xu, B.H.; Xia, Q.; Sun, B.; Ren, Y.; Wang, Z.L. A CNN-SAM-LSTM hybrid neural network for multi-state estimation of lithium-ion batteries under dynamical operating conditions. *Energy* **2024**, *294*, 130764. [CrossRef]
25. Zhang, Y.; Wang, Y.Q.; Zhang, C.; Qiao, X.J.; Ge, Y.D.; Li, X.; Peng, T.; Nazir, M.S. State-of-health estimation for lithium-ion battery via an evolutionary Stacking ensemble learning paradigm of random vector functional link and active-state-tracking long–short-term memory neural network. *Appl. Energy* **2024**, *356*, 122417. [CrossRef]
26. Sui, X.; He, S.; Vilsen, S.B.; Meng, J.H.; Teodorescu, R.; Stroe, D.I. A review of non-probabilistic machine learning-based state of health estimation techniques for Lithium-ion battery. *Appl. Energy* **2021**, *300*, 117346. [CrossRef]
27. Li, Y.Y.; Sheng, H.M.; Cheng, Y.H.; Stroe, D.I.; Teodorescu, R. State-of-health estimation of lithium-ion batteries based on semi-supervised transfer component analysis. *Appl. Energy* **2020**, *277*, 115504. [CrossRef]
28. Zhang, L.S.; Wang, W.T.; Yu, H.Q.; Zhang, Z.; Yang, X.B.; Liang, F.W.; Li, S.; Yang, S.C.; Liu, X.H. Remaining useful life and state of health prediction for lithium batteries based on differential thermal voltammetry and a deep learning model. *iScience* **2022**, *12*, 105638. [CrossRef]

batteries

MDPI

Article

State of Health Estimation of Li-Ion Battery via Incremental Capacity Analysis and Internal Resistance Identification Based on Kolmogorov–Arnold Networks

Jun Peng [1], Xuan Zhao [1,*], Jian Ma [1], Dean Meng [1], Shuhai Jia [2], Kai Zhang [1], Chenyan Gu [1] and Wenhao Ding [1]

1 School of Automobile, Chang'an University, Xi'an 710064, China; pengjun@chd.edu.cn (J.P.)
2 School of Mechanical Engineering, Xi'an Jiaotong University, Xi'an 710049, China
* Correspondence: zhaoxuan@chd.edu.cn

Abstract: An accurate estimation of the state of health (SOH) of Li-ion batteries is critical for the efficient and safe operation of battery-powered systems. Traditional methods for SOH estimation, such as Coulomb counting, often struggle with sensitivity to measurement noise and time-consuming tests. This study addresses this issue by combining incremental capacity (IC) analysis and a novel neural network, Kolmogorov–Arnold Networks (KANs). Fifteen features were extracted from IC curves and a 2RC equivalent circuit model was used to identify the internal resistance of batteries. Recursive least squares were used to identify the parameters of the equivalent circuit model. IC features and internal resistance were considered as input variables to establish the SOH estimation model. Three commonly used machine learning methods (BP, LSTM, TCN) and two hybrid algorithms (LSTM-KAN and TCN-KAN) were used to establish the SOH estimation model. The performance of the five models was compared and analyzed. The results demonstrated that the hybrid models integrated with the KAN performed better than the conventional models, and the LSTM-KAN model had higher estimation accuracy than that of the other models. The model achieved a mean absolute error of less than 0.412% in SOH prediction in the test and validation dataset. The proposed model does not require complete charge and discharge data, which provides a promising tool for the accurate monitoring and fast detection of battery SOH.

Keywords: SOH estimation; KAN; incremental capacity analysis; internal resistance; Li-ion battery

1. Introduction

Electric and hybrid vehicles have experienced rapid growth in recent years. Lithium-ion (Li-ion) batteries are a critical component of electric vehicles (EVs) in providing power for vehicles. The application scale of Li-ion batteries in EVs is far larger than other types of batteries owing to several advantages of Li-ion batteries, including high energy density, a long lifespan, low maintenance requirements, etc. [1]. A battery management system (BMS) plays a key role in ensuring the safe and efficient operation of a vehicle's battery system. State of health (SOH) estimation is an indispensable function for the BMS, which contributes to determining retained capacity and remaining useful life, optimizing the charging/discharging process to extend the life and prevent the premature failure of batteries [2].

While the SOH is crucial for a battery management system (BMS) to function effectively, an accurate definition and calculation method for the SOH has not been universally established [3]. Generally, capacity fading and the increase in internal resistance are commonly used indicators to assess the SOH of a battery. To obtain variation in capacity and resistance, the most commonly used approaches are to measure the current and voltage parameters of a battery to derive the two indicators. The approaches to acquiring the SOH from the current and voltage data can be categorized into two types: direct computation [4] or model-based machine learning [5]. The former method includes coulomb counting, internal resistance/impedance calculation, etc. The method based on machine learning uses the current/voltage data or other variables derived from them to establish a neural network model to predict the change in capacity and impedance.

Coulomb counting is an integral of current over time that uses a current sensor to monitor the current during the charge/discharge process, and the maximum increased/elapsed capacity can be calculated. The SOH of batteries can be determined through the ratio of maximum capacity to the rated capacity [6]. Battery SOH can also be defined from the perspective of internal resistance evolution. The internal resistance increases with the degeneration of cells. The difference between the internal resistance at the end of battery life and the fresh stage is regarded as a basis for evaluating the SOH [7]. This resistance, R, can be obtained by Ohm's law or parameter identification based on an equivalent circuit model.

These above approaches have their limitations. For instance, the Coulomb counting method is only available in a full charge/discharge process to obtain the current maximum capacity. The accuracy of resistance measurement is influenced by the battery state of charge and the environmental temperature. Therefore, the SOH estimation method based on the calculation of capacity and resistance is restricted by the working circumstance of batteries and has low efficiency and poor scalability. To overcome this challenge, data-driven approaches are proposed to predict the SOH based on the history data of the current, voltage, capacity, etc., wherein the statistical analysis method and machine learning are mostly used in the establishment of SOH estimation models, such as gray relation analysis [8], support vector regression (SVR) [9], XGboost [10], back propagation (BP) neural network [11,12], and Long Short-Term Memory (LSTM) network [13,14]. In May 2024, a team of researchers at the Massachusetts Institute of Technology introduced the Kolmogorov–Arnold Network (KAN), a novel neural network architecture differing from the multilayer perceptron (MLP) network [15]. This innovative approach is grounded in the theoretical principles of the Kolmogorov–Arnold representation theorem, which offers a powerful framework for decomposing complex multivariate functions into simpler, univariate components [16].

The key steps of data-driven approaches are defining effective informative inputs and establishing a robust estimation model. Commonly, the current, voltage, and temperature data in the charge/discharge process are set as the inputs. Compared to the unprocessed current and voltage data, the features derived from them can better describe the battery aging mechanism and understand the materials' properties, interfacial phenomena, and electrochemical behavior inside cells, such as electrochemical impedance spectroscopy (EIS), incremental capacity (IC), and differential voltage (DV). Incremental capacity analysis is a widely used approach to identify and quantify the changes in electrochemical reactions inside a battery [17–19]. It is assumed that the cell is at an equilibrium state during a charge/discharge process with an extremely low current rate (C-rate). However, the low C-rate charge/discharge process cannot be performed in field application due to the high cost of time. For SOH estimation, it is unnecessary to pay more attention to the details of the battery degradation process; rather, it is necessary to just focus on the remaining battery capacity at a certain moment. Therefore, a pragmatic solution is proposed to replace a low C-rate with a large C-rate. Therefore, there is increasing research on using IC curve data for SOH prediction. Li et al. adopted the Gaussian filter to process IC curves and use the position and height of the curve peak as the feature of interest for the SOH estimation of a high-energy NMC li-ion battery [20]. In ref. [21], IC curves were used to estimate battery SOH based on SVR. Beltran et al. investigated the performance of different machine learning methods in the estimation of retained capacity, using IC curves [22].

The internal resistance of a battery comprises several components that collectively determine how much opposition the battery presents to the flow of the electric current. These components can be broadly categorized into three main types: ohmic resistance, polarization internal resistance, and electrochemical impedance [23]. Ohmic resistance is the straightforward electrical resistance of the battery's conductive materials (electrodes, electrolytes, and connectors) [24]. Electrochemical impedance is more complex and includes various factors associated with the battery's electrochemical processes, including charge transfer resistance, and solid electrolyte interphase (SEI) resistance [25]. As a battery undergoes repeated charge and discharge cycles, various degradation will occur, such as the growth of the SEI layer, electrolyte decomposition, and loss of active material. These processes lead to an increase in the internal resistance of the battery. An increase in internal resistance is a key indicator of these degradation processes [26,27]. It reflects the battery's reduced efficiency in conducting current, which is directly related to its aging and SOH. Internal resistance is a valuable and reliable indicator for estimating the SOH of a battery [28].

In the above works, only IC curve features were used as input parameters for modeling. IC curve features are susceptible to the environmental temperature and current rate. In this work, we employed both internal resistance parameters and IC features as inputs to establish a hybrid model. Moreover, the aforementioned machine learning models were based on MLP. KANs, as an emerging network structure, have the potential to become an alternative to MLP. Therefore, we proposed an SOH estimation approach by the combination of features extracted from IC curves and internal resistance based on the KAN algorithm. First, an accelerated aging test was performed to prepare the degradation data. Next, the raw data of charge cycles were processed to obtain the IC curves, and some advanced filter algorithms were adopted to smooth the original IC curves. The peaks of IC curves often correspond to phase transitions within the battery's active materials. For example, in lithium-ion batteries, lithium intercalates into or de-intercalates from electrode materials. In some cases, an IC curve peak might indicate lithium plating, which is a form of battery degradation [29,30]. The dips represent gaps between different intercalation stages within the battery's active materials. The peaks and dips of an IC curve of a battery provide important insights into the battery's state and characteristics. Therefore, the features of IC curves were extracted as the inputs of SOH estimation models, including the height, width, and position of peaks and dips. Then, a 2RC equivalent circuit model and the recursive least squares (RLS) algorithm were used to identify the internal resistance of batteries. Three commonly used machine learning methods were used to establish the SOH estimation model, including BP, LSTM, and Temporal Convolutional Networks (TCNs), and the KAN was adopted to build hybrid models (LSTM-KAN and TCN-KAN) for SOH estimation. Furthermore, we compared the estimation performance of five models and investigated the effect of adding the KAN on the original model performance. The model with the best performance among the five was chosen to evaluate the accuracy of the SOH estimation model with different input variables. To our best knowledge, this was the first time the KAN was used to estimate the SOH of Li-ion batteries.

The rest of this paper is organized as follows. In Section 2, we describe the experiment and data preparation in detail, including the battery aging test conditions, the feature extraction from the IC curve, and internal resistance identification. The principle of LSTM-KAN and the SOH estimation model framework based on LSTM-KAN are shown in Section 3. The prediction results of the proposed SOH estimation model and the comparison between different models and input parameters are shown in Section 4, and the conclusions are drawn in Section 5.

2. Experiment and Data Preparation

2.1. Battery Aging Test

To obtain the aging data of batteries, a test platform was established, as shown in Figure 1. A battery testing system, BTS4000 (Neware, Shenzhen, China), was used to run the defined charge/discharge cycles. Eight commercial $LiFePO_4$ batteries were tested in the experiment, and cell specification is shown in Table 1. The cells were placed in the temperature control chamber to maintain a stable ambient temperature during the charging/discharging process. Capacity degeneration is closely related to the charging environmental temperature. To accelerate the aging process, the ambient temperature for charging and discharging was set at 35 °C. A standard charge/discharge regime was applied on the cells, which was composed of constant current (CC) charging with 1C to the cut-off voltage, then constant voltage charging to the cut-off current of C/20, and CC discharging with 1C to a cut-off voltage. The rest time between charge and discharge was set to 0.5 h.

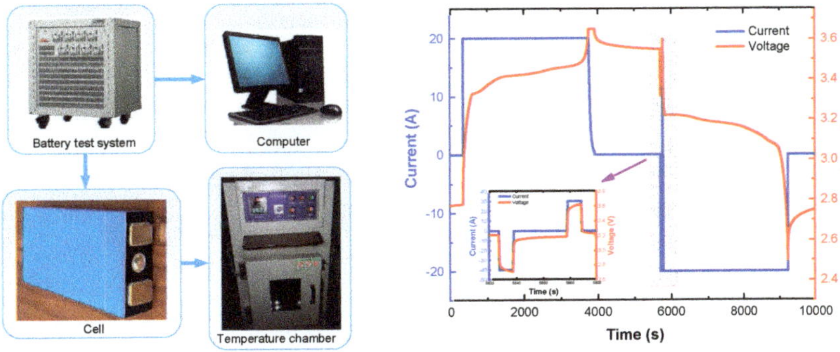

Figure 1. Experiment setup for the aging test.

Table 1. Specification of cells used in the experiment.

Item	
Nominal capacity	20 Ah
Nominal voltage	3.2 V
Charging cut-off voltage	3.65 V
Discharge cut-off voltage	2.5 V
Dimension	$133 \times 70 \times 28$ mm^3
Temperature	35 °C
Sampling frequency	1 s

2.2. IC Curve Acquisition and Feature Extraction

Incremental capacity analysis is a widely used technique to obtain information on the electrochemical properties of a cell and unveil degradation mechanisms. IC curves describe the increment of battery capacity within an interval of voltage. The calculation formula is given in Equation (3).

$$IC = \frac{dQ}{dV} = \frac{\Delta Q}{\Delta V} \tag{1}$$

where ΔQ is the increment of capacity in the charge/discharge cycles. ΔV is the corresponding voltage interval.

The IC curve reveals the phase-equilibrium-induced voltage plateau. The peaks of curves are in the range of the voltage plateau. In the voltage plateau region, the capacity increases rapidly with small voltage changes. Due to the limited voltage sampling frequency of charging and discharging equipment, the linear interpolation method was used to obtain the capacity with equal voltage intervals. A large voltage interval may miss some significant information in this region. To obtain complete information on the IC curve as soon as possible, the small voltage internal was set as 1 mV. The peak value of the IC curve and its corresponding voltage platform are shown in Figure 2. The slope of the voltage capacity (V-Q) curve has three obvious changes, which correspond to the three conspicuous peaks of the IC curve. Each peak represents the phase transition point of the materials inside cells, which can characterize the lithiation plateau, transforming into different lithiation stages. The area enclosed by the curve and axis manifests the capacity change in the phase transition. From the picture, it can be observed that the raw IC in the range of the voltage plateau fluctuates greatly. Therefore, it is necessary to smooth the original curves through a filtering algorithm. In this work, the Gaussian filter method was adopted to preprocess the raw IC curves, which can alleviate the effect of the interference in different cycles on significant IC curve feature extractions.

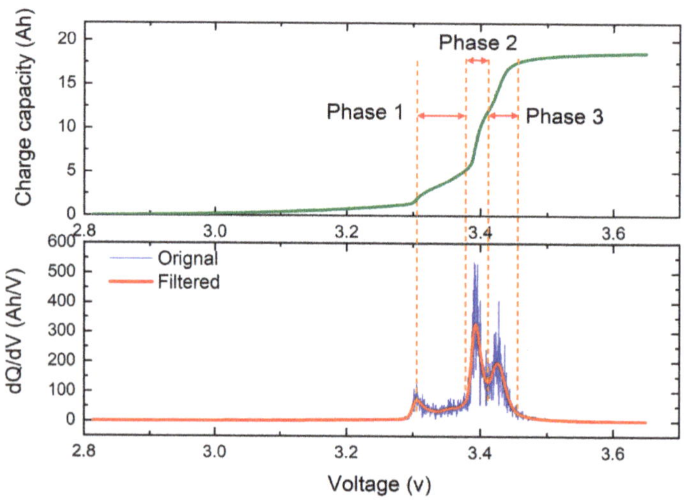

Figure 2. Capacity variation with voltage and IC curve during a charging cycle.

To investigate the relationship between the charge capacity and voltage, partial V-Q curves during 700 cycles are demonstrated in Figure 3. The curves almost overlap in the range from 2.7 V to 3.15 V. In the range of 3.15 V–3.65 V, the difference between curves is relatively large due to capacity degeneration. From this view, as the cycle increases, the position of the voltage plateau is offset to the right. This phenomenon suggests that the peaks of IC curves shift as the battery ages. However, as the cycle progresses, this trend becomes less significant. From the enlarged view, the slope of the voltage plateau becomes slowed down from cycle 1 to cycle 701, especially in phase 3. In other words, the heights of peaks decrease with battery degradation.

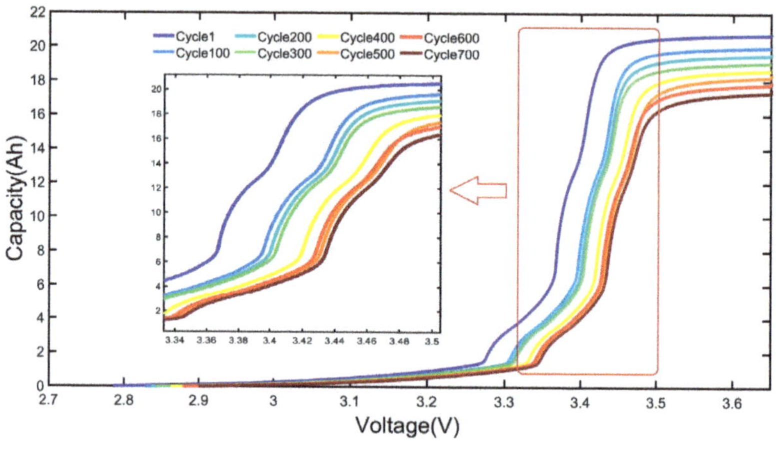

Figure 3. V-Q curves of battery in different aging states during a charge cycle.

To further verify these findings, all the IC curves during 701 cycles are demonstrated in Figure 4. It is evident that the positions of curve peaks shift and the height of the peaks changes, especially the third peak. Moreover, the initial and end voltage values of the voltage platform are also different. As a battery ages, the peaks may become less pronounced, and the IC curve may flatten overall. This demonstrates that changes in the positions, shapes, and heights of peaks and dips over time can provide information about a battery's state of health. Therefore, we extracted the features of IC curves considering these changes, and the features include the positions of peaks (L_{P1}, L_{P2}, L_{P3}), heights of peaks (H_{P1}, H_{P2}, H_{P3}), width of peaks (W_{P1}, W_{P2}, W_{P3}), positions of dips (L_{d1}, L_{d2}, L_{d3}), and heights of dips (H_{d1}, H_{d2}, H_{d3}), as shown in Figure 5. It was observed that all the peak features were in the voltage range of 3.2 V to 3.5 V. The state of charge of batteries in the voltage range was 20–85%, which is exactly the actual working range of most battery systems such as electric vehicles.

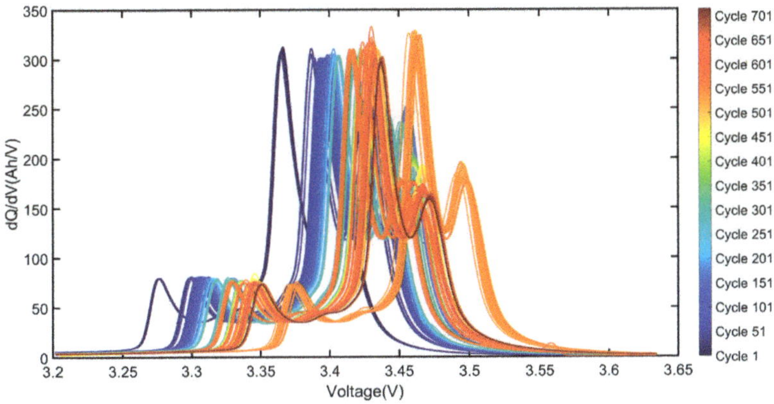

Figure 4. Variations in IC curves during 700 cycles.

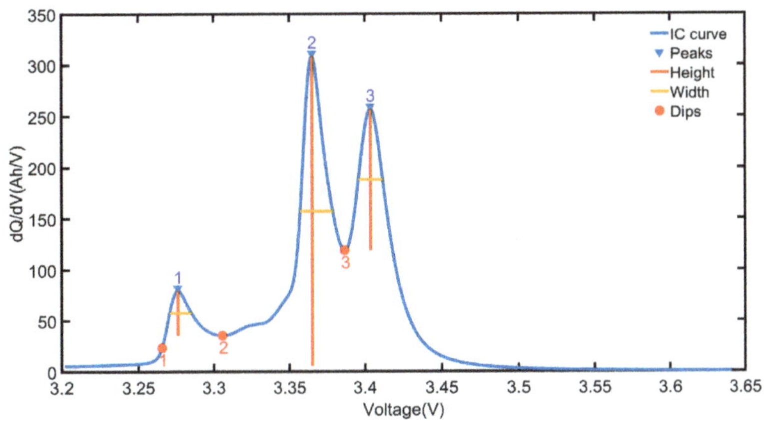

Figure 5. Features extracted from the IC curve.

2.3. Internal Resistance Identification

2.3.1. Model Establishment

Equivalent circuit models are widely used to recognize and estimate the internal parameters of batteries, which are crucial for understanding and managing battery performance. These models provide a simplified representation of the complex electrochemical processes occurring within the battery, making it easier to analyze and control. Commonly used equivalent circuit models of batteries include Rint, Thevenin, 2RC, PNGV, and GNL models. The Rint model is simple, involving only an OCV and a single resistor, which does not capture the dynamic response of the battery to changing loads and conditions, leading to less accurate predictions of battery performance. The single RC branch in the Thevenin model captures only basic transient behavior, leading to less accurate representations of the battery's dynamic performance, especially under varying load conditions. The PNGV and GNL models are more sophisticated battery models, so the process of identifying and calibrating the parameters is more complex and time-consuming compared to the 2RC model. The 2RC model provides a more accurate representation of a battery's behavior, especially under varying load conditions. By including two RC pairs, this model captures both the short-term and long-term dynamics of the battery, reflecting its real-world performance more closely. Compared to the Thevenin model, the 2RC model can more effectively represent these non-linearities, improving the predictive capability of simulations and control systems. By capturing both fast and slow dynamics, the second-order RC model can provide better insight into the SOH of the battery. This allows for more precise monitoring and diagnostics of parameters like internal resistance and capacitance, which are indicative of aging and degradation. A 2RC equivalent circuit model was used in this work, as shown in Figure 6. R_0, R_1, and R_2 represent the ohmic internal resistance and polarization resistance of cells, respectively. The dynamics of 2RC circuits can be given as shown in Equation (2) [31].

$$\begin{aligned} U_{oc}(t) - U_L(t) &= U_{P1}(t) + U_{P2}(t) + i(t)R_0 \\ \dot{U}_1(t) &= \frac{i(t)}{C_1} - \frac{U_1}{C_1R_1} \\ \dot{U}_2(t) &= \frac{i(t)}{C_2} - \frac{U_2}{C_2R_1} \end{aligned} \quad (2)$$

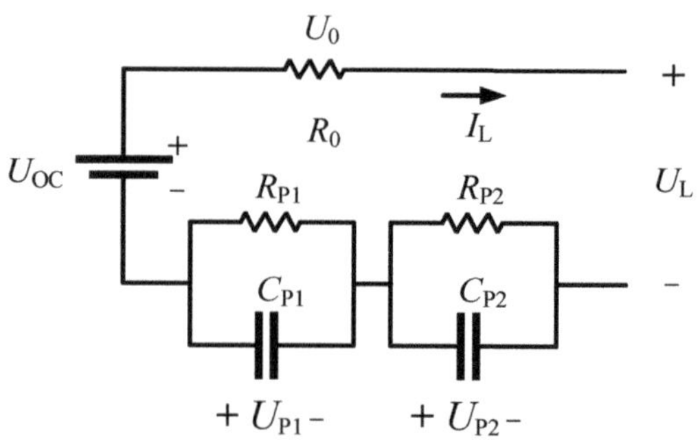

Figure 6. RC equivalent circuit model of batteries.

According to Equation (2), the complex frequency domain dynamics model of the 2RC circuit can be given in

$$U_{oc}(s) - U_L(s) = I(s)\left(R_0 + \frac{R_1}{R_1C_1s + 1} + \frac{R_2}{R_2C_2s + 1}\right) \quad (3)$$

The transfer function is expressed as

$$G(s) = \frac{U_{oc}(s) - U_L(s)}{I(s)} = \left(R_0 + \frac{R_1}{R_1C_1s + 1} + \frac{R_2}{R_2C_2s + 1}\right) \quad (4)$$

The z-transform can be used to discretize the above continuous equation by the following equation:

$$s = \frac{2(1 - z^{-1})}{T(1 - z^{-1})} \quad (5)$$

Through a combination of Equations (4) and (5), the discrete transferring function can be expressed as

$$G(z^{-1}) = \frac{a_3 + a_4z^{-1} + a_5z^{-2}}{1 - a_1z^{-1} - a_2z^{-2}} \quad (6)$$

$$\begin{cases} a_1 = \frac{8\tau_1\tau_2 - 2T^2}{T^2 + 4\tau_1\tau_2 + 2T\tau_1 + 2T\tau_2} \\ a_2 = \frac{2T\tau_1 + 2T\tau_2 - T^2 - 4\tau_1\tau_2}{T^2 + 4\tau_1\tau_2 + 2T\tau_1 + 2T\tau_2} \\ a_3 = \frac{T^2R_1 + T^2R_2 + 2TR_1\tau_2 + 2TR_2\tau_1}{T^2 + 4\tau_1\tau_2 + 2T\tau_1 + 2T\tau_2} \\ a_4 = \frac{2T^2R_1 + 2T^2R_2}{T^2 + 4\tau_1\tau_2 + 2T\tau_1 + 2T\tau_2} \\ a_5 = \frac{T^2R_1 + T^2R_2 - 2TR_1\tau_2 - 2TR_2\tau_1}{T^2 + 4\tau_1\tau_2 + 2T\tau_1 + 2T\tau_2} \end{cases} \tag{7}$$

where T is the sampling interval. A_1, a_2, a_3, a_4, and a_5 are the coefficients related to the model parameters. $\tau_1 = R_1C_1$, and $\tau_2 = R_2C_2$.

The differential equation of the RC model is given as

$$U(k) = a_1U(k-1) + a_2U(k-2) + a_3I(k) + a_4I(k-1) + a_5I(k-2) \tag{8}$$

The RLS algorithm is extensively employed in battery parameter identification due to its efficiency and accuracy in the real-time estimation of dynamic systems. The RLS algorithm is an adaptive filter algorithm that recursively finds the coefficients that minimize a weighted linear least squares cost function relating to the input signals. It is well suited for systems where parameters change over time, making it ideal for battery modeling. The model can be described by the recursive process of the RLS algorithm as follows:

$$\begin{array}{l} U(k) = \varphi^T(k)\theta(k) \\ \varphi(k) = [U(k-1), U(k-2), I(k), I(k-1), I(k-2)]^T \\ \theta(k) = [a_1, a_2, a_3, a_4, a_5]^T \end{array} \tag{9}$$

Combining the above equations, the parameters can be expressed as

$$\begin{array}{l} R_0 = [a_3 - a_4 + a_5]/[1 + a_1 - a_2] \\ R_1C_1R_2C_2 = [T^2(1 + a_1 - a_2)]/[4(1 - a_1 - a_2)] \\ R_1C_1 + R_2C_2 = [T(1 + a_2)]/[1 - a_1 - a_2] \\ R_0 + R_1 + R_2 = [a_3 - a_4 + a_5]/[1 - a_1 - a_2] \\ R_0R_1C_2 + R_0R_2C_2 + R_1R_2C_1 + R_1R_2C_2 = T[a_5 - a_3]/[1 - a_1 - a_2] \end{array} \tag{10}$$

The parameters of the equivalent circuit model can be determined by calculating Equation (10).

2.3.2. Identification Results

The Hybrid Pulse Power Characterization (HPPC) test profile was added in each discharge cycle, as demonstrated in Figure 4, and was used to generate data for identifying the parameters of 2RC models. Internal resistance varies with the environmental temperature, the SOC of batteries, and the SOH. To eliminate the interference of temperature and the SOC, testing was carried out in a constant environment temperature, and the HPPC test was conducted in the same SOC state during the discharge phase. The data of HPPC were used to identify the parameters of the 2RC model by the RLS algorithm, as shown in Figure 7. As demonstrated in Figure 7a, the predicted terminal voltage was in good agreement with the experimental one. The error between the predicted and experimental is illustrated in Figure 7b, and the mean absolute error was less than 0.4 mV. This indicates that the identified parameters accurately represented the battery's behavior. The identified parameters of the equivalent model are shown in Figure 7c,d. The average value of the identification parameters during the relaxation phase was taken as the final value for each parameter.

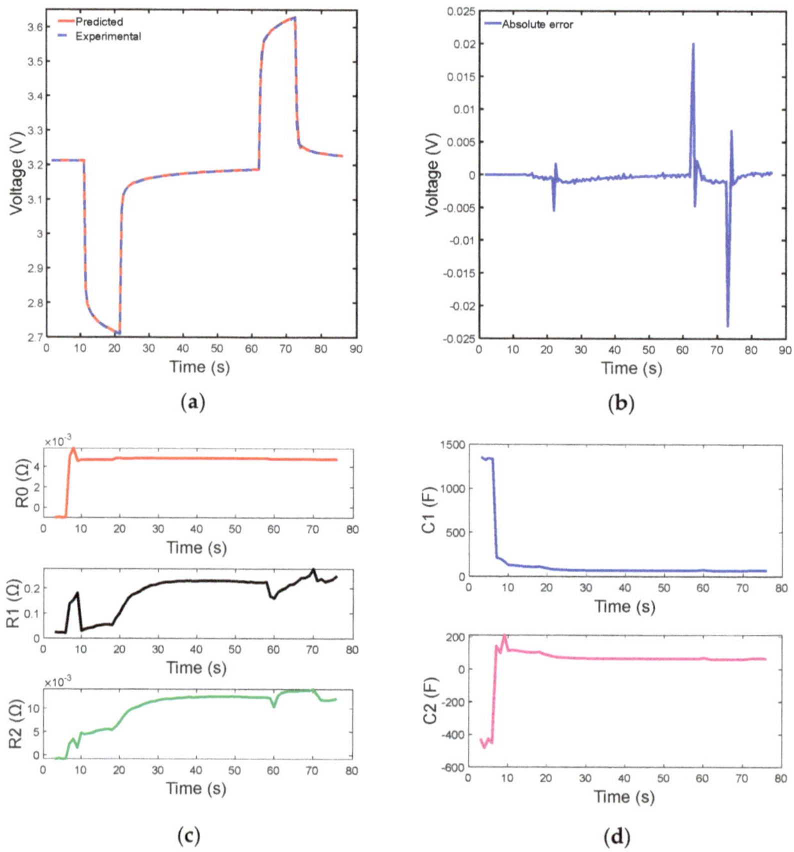

Figure 7. Results of parameter identification for the 2RC equivalent circuit model. (**a**) the comparison of simulated and experimental voltage; (**b**) the simulation error; (**c**) and (**d**) the identification value of R_0, R_1, R_2, C_1, C_2.

3. SOH Estimation Based on LSTM-KAN

3.1. LSTM-KAN Architecture

A Long Short-Term Memory Network is a special type of Recurrent Neural Network (RNN) capable of learning long-term dependencies, which has since been widely adopted for various tasks involving sequential data. The core component of an LSTM network is the memory cell, which can maintain its state over time. This allows the network to remember important information for long periods. LSTMs have three types of gates that regulate the flow of information, including forget gate, input gate, and output gate. The forget gate decides which parts of the cell state to forget. It takes the previous hidden state (h_{t-1}) and the current input (x_t) and passes them through a sigmoid function, as follows:

$$f_t = \sigma(W_f \cdot [h_{t-1}, x_t] + b_f) \tag{11}$$

The input gate decides which new information to store in the cell state. It has two parts: the input gate layer and candidate values. The former determines which values to update, and the latter creates new candidate values to be added to the state. The formulas of the input gate layer and candidate values can be described as follows:

$$i_t = \sigma(W_i \cdot [h_{t-1}, x_t] + b_i) \tag{12}$$

$$\widetilde{C}_n = \tanh(W_c \cdot [h_{t-1}, x_t] + b_c) \tag{13}$$

The new cell state is a combination of the old state and the new candidate values.

$$C_t = f_t \cdot C_{t-1} + i_t \cdot \widetilde{C}_t \tag{14}$$

The output gate determines the next hidden state, which is a filtered version of the cell state.

$$O_t = \sigma(W_o \cdot [h_{t-1}, x_t] + b_o) \tag{15}$$

The hidden state output of the network node is described and calculated by this equation, as follows:

$$h_t = O_t \tanh(C_t) \tag{16}$$

Kolmogorov–Arnold Networks (KANs) are a novel network architecture proposed by the MIT team in May 2024, which were inspired by the Kolmogorov–Arnold representation theorem [15], as shown in Figure 8. This theorem states that any multivariable continuous function can be decomposed into a finite sum of continuous functions of one variable and an additional continuous function. Leveraging this theoretical foundation, Kolmogorov–Arnold Networks aim to simplify the representation of complex functions and improve the efficiency and interpretability of neural networks.

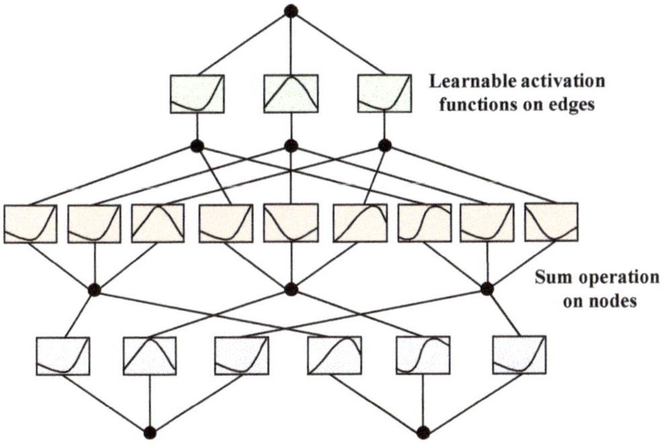

Figure 8. Kolmogorov–Arnold Network architecture.

The architecture of KANs typically involves a decomposition of the input space into individual dimensions, followed by processing through one-dimensional functions before combining the results. The theorem can be expressed as follows:

$$f(x_1, \ldots, x_n) = \sum_{q=1}^{2n+1} \Phi_q \left(\sum_{p=1}^{n} \varphi_{q,p}(x_p) \right) \tag{17}$$

where $\varphi_{q,p}$ are univariate functions that map each input variable x_p, and Φ_q are continuous functions. This enables a KAN to model complex interactions in high-dimensional data by breaking them down into compositions of simpler univariate functions.

A KAN layer with n_{in}-dimensional inputs and n_{out}-dimensional outputs can be defined as a matrix of 1D functions.

$$\Phi = \{\phi_{q,p}\}, p = 1, 2, \ldots, n_{\text{in}}, q = 1, 2, \ldots, n_{\text{out}} \tag{18}$$

where the $\phi_{q,p}$ is parametrized functions with l trainable parameter.

A deeper KAN is composed of multiple KAN layers with a strong ability to model more complex functions. The architecture of a deeper KAN can be expressed as follows:

$$KAN(x) = (\Phi_{L-1} \circ \Phi_{L-2} \circ \ldots \circ \Phi_0)(x) \tag{19}$$

LSTM-KAN combines the strengths of LSTM networks with the theoretical insights of Kolmogorov–Arnold Networks to model complex, time-dependent relationships in high-dimensional data. This hybrid approach leverages an LSTM's ability to handle sequential data and a KAN's capacity for simplifying multivariate functions into compositions of univariate functions. The hybrid network is implemented by integrating a KAN directly into the hidden layer of an LSTM network. This can be especially useful for applications requiring both temporal sequence analysis and complex feature interactions.

3.2. SOH Estimation Model Framework

In the above section, the features of IC curves and the parameters of internal resistance were obtained. Next, the extracted features and internal resistance were taken as input parameters, and the output was the SOH. From Figure 4, it is observed that the IC curves did not all shift to the right as the cycle increased; rather, the curve offset presented a time series characteristic. Here, the LSTM network was used to process the IC curve features and internal resistance parameters and capture the temporal dependencies. In the hidden layer before output, the KAN layer was used to replace the standard fully connected layers in the LSTM. The structure of the proposed SOH estimation model based on LSTM-KAN is shown in Figure 9. This framework provides a robust approach to estimate the battery SOH using LSTM-KAN and features derived from IC curves, leveraging the strengths of the LSTM in handling sequential data and the KAN in decomposing complex multivariate functions to achieve accurate and reliable SOH predictions. The input parameter matrix consists of 18 parameters, including 15 IC features, and 5 parameters of 2RC equivalent circuit models.

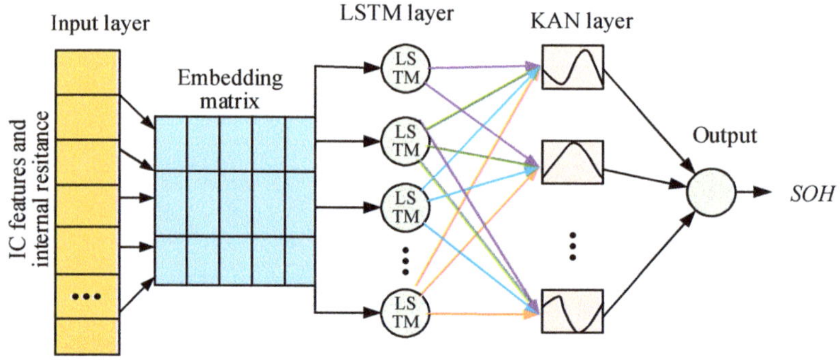

Figure 9. Structure of LSTM-KAN used for SOH estimation.

3.3. Evaluation of the Error of the SOH

Evaluating the error of SOH estimation models is critical in order to ensure their accuracy and reliability. Three error metrics are utilized to evaluate the performance of different estimation models.

Mean Absolute Error (MAE)

$$\text{MAE} = \frac{1}{n}\sum_{i=1}^{n}\left|y_i - \widehat{y}_i\right| \tag{20}$$

Mean Absolute Error (MAPE)

$$\text{MAPE} = \frac{1}{n}\sum_{i=1}^{n}\left|\frac{y_i - \widehat{y}_i}{y_i}\right| * 100\% \tag{21}$$

Mean Absolute Error (RMSE)

$$\text{RMSE} = \sqrt{\frac{1}{n}\sum_{i=1}^{n}(y_i - \widehat{y}_i)^2} \tag{22}$$

where y_i and $\widehat{y}_i$ are the real value and predicted value at test cycle I, and n is the total of the test cycles.

4. Results of SOH Estimation

4.1. The SOH Estimated by the LSTM-KAN Model

The data used in this experiment come from the accelerated aging test, and eight batteries with different initial SOHs were selected as the samples. Four fresh cells and four cells with the same specifications were selected for the aging tests. The initial SOH and aging cycles experienced are shown in Table 2. Selecting batteries with different SOH values not only enriches the sample types in the training data but can also be used to verify the generalization ability of the estimation model under different SOH conditions.

Table 2. SOH initial values and the experienced cycles of battery samples.

	Bat. #1	Bat. #2	Bat. #3	Bat. #4	Bat. #5	Bat. #6	Bat. #7	Bat. #8
SOH	100%	100%	100%	100%	90.3%	87.9%	88.09%	96.18%
Cycles	500	1070	930	380	550	375	320	590

The training dataset is composed of the IC features and parameters of the equivalent circuit of six batteries (Bat. #1, #2, #3, #6, #7, and #8). The data of Bat. #4 and Bat. #8 are used as the test and validation dataset, respectively. The SOH of Bat. #4 was estimated using the three most commonly used machine learning methods and the hybrid models based on the KAN, as shown in Figure 10. The actual SOH shows a steady decline over the cycle range, from 0 to 600. The SOH value predicted by the five models follows the general trend in the actual SOH but with different degrees of fluctuation. From the picture, the SOH predicted by the BP model has significant deviation from the actual value. The values estimated by the LSTM-KAN and TCN-KAN models are better matched with the actual SOH. The results indicate that while the conventional models can capture the overall degradation trend, they show some noise or variability in their predictions. The SOH predicted by hybrid models closely follows the actual value with less fluctuation.

To ensure the trained model's robustness and generalization ability, a validation dataset was used to evaluate its performance. The data for Bat. #5 were used as the validation dataset. The results of the validation are illustrated in Figure 11, which demonstrates that the trained models also have good performance on the validation dataset. The values predicted by the five models show a similar trend to the actual SOH but with a noticeable offset from the actual SOH in the BP and TCN models. From the enlarged view, it can be observed that the SOH predicted by the LSTM-KAN model is closer to the actual value. Similar to the results of the test data, the prediction error of the BP model is the largest in the validation dataset.

The results of the test and validation data confirm the superior performance and generalization ability of the hybrid model compared to the standalone network. The consistency of the hybrid model's performance across both the training and validation datasets suggests that it effectively captures the underlying patterns of battery degradation. This demonstrates that the hybrid model not only performs well on the training data but also maintains high accuracy and stability when applied to new data. The error metrics of the five models on the test and validation dataset are shown in Table 3. It demonstrates that the prediction errors of the BP model are the largest among the five models. All the errors of the hybrid model are smaller than those of the original model. In the test dataset, the LSTM-KAN model has the fewest errors, and the MAE, MAPE, and RMSE are 0.412%, 0.462%, and 0.570%, respectively. In the validation dataset, the MAE of the LSTM and LSTM-KAN are 0.438% and 0.256%, which is in good agreement with the results of the test data. The SOH estimation performance of the LSTM-KAN and TCN-KAN models is comparable. The above results demonstrate that combining the KAN can effectively improve the estimation performance of conventional machine learning models, and the LSTM-KAN model has the best performance in SOH estimation, with a comprehensive comparison of testing and validation data results. Therefore, the LSTM-KAN model was used to further investigate the model performance under different input parameters. The training time of each model is shown in Figure 12. In Refs. [11,12], the SOH estimation models are established based on similar methods, but the estimation error (MAE: 1.16%, 1.957%) of the model in the literature is far larger than that proposed in this work (MAE: 0.412%).

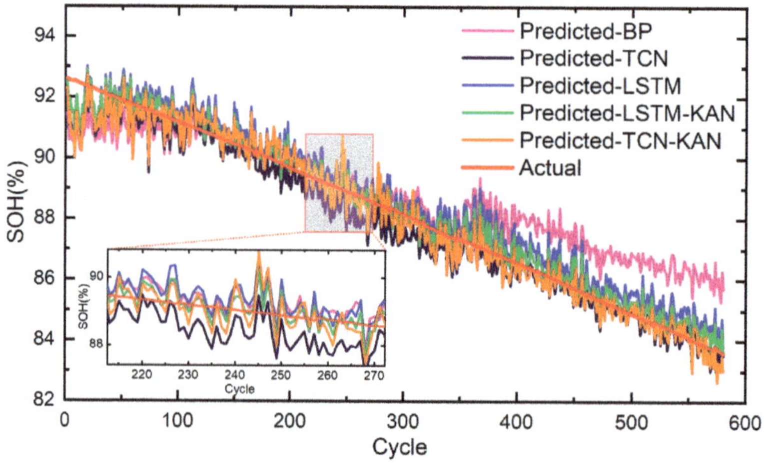

Figure 10. Predicted SOH of Bat. #4 for the test dataset.

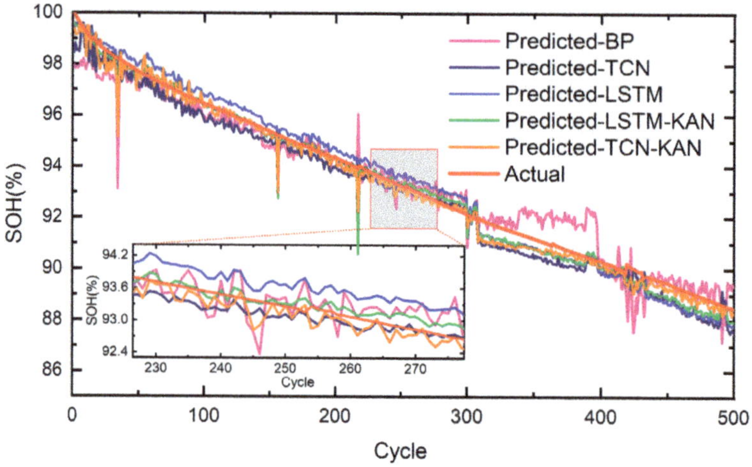

Figure 11. SOH of Bat. #5 predicted by the trained model.

Table 3. SOH estimation errors of two models in different datasets.

	MAE	MAPE	RMSE
Test-BP	0.811%	0.921%	1.028%
Test-TCN	0.472%	0.528%	0.607%
Test-LSTM	0.519%	0.585%	0.641%
Test-LSTM-KAN	0.412%	0.462%	0.570%
Test-TCN-KAN	0.426%	0.476%	0.572%
Validation-BP	0.531%	0.569%	0.727%
Validation-TCN	0.461%	0.495%	0.547%
Validation-LSTM	0.438%	0.472%	0.490%
Validation-LSTM-KAN	0.256%	0.278%	0.378%
Validation-TCN-KAN	0.251%	0.268%	0.365%

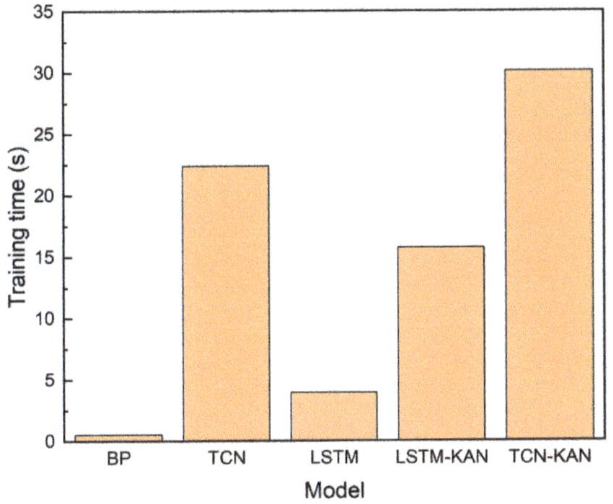

Figure 12. Training time of each model.

4.2. The SOH Estimated by the LSTM-KAN Model with Different Input Variables

To investigate the improvement of the model after integrating internal resistance parameters into the input variables, the training data were divided into two groups: group I and group II. The input variables in model I include 15 features extracted from IC curves and internal resistance parameters, and those in model II only contain 15 features extracted from IC curves, as demonstrated in Table 4. The corresponding SOH estimation models, model I and model II, were trained using datasets from group I and group II, respectively. Figure 13 illustrates the SOH of Bat. #5 predicted by two models. The results show that the predicted SOH is closely aligned with the actual value. Model II shows greater variability and deviations from the actual SOH compared to model I. The performance of the two models on the validation samples is demonstrated in Figure 14. Similarly, model I demonstrates better prediction performance than that of model II. From the subplots, model I has a lower MAE in the test and validation dataset. This indicates that the integration of internal resistance parameters into the input parameters can improve the model's performance.

Table 4. Input and output variables of model I and model II.

	Input	Output
Model I	$L_{P1}, L_{P2}, L_{P3}, H_{P1}, H_{P2}, H_{P3}, W_{P1}, W_{P2}, W_{P3}, L_{d1}, L_{d2}, L_{d3}, H_{d1}, H_{d2}, H_{d3}, R_0, R_1, R_2, C_1, C_2$	SOH
Model II	$L_{P1}, L_{P2}, L_{P3}, H_{P1}, H_{P2}, H_{P3}, W_{P1}, W_{P2}, W_{P3}, L_{d1}, L_{d2}, L_{d3}, H_{d1}, H_{d2}, H_{d3}$	SOH

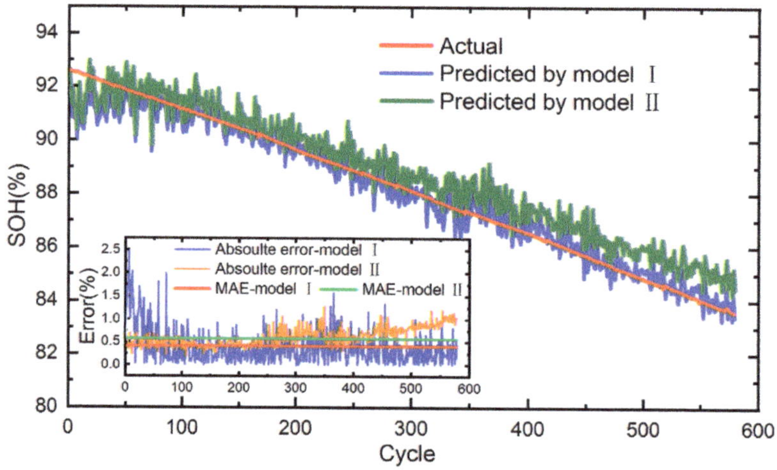

Figure 13. SOH of Bat. #4 predicted by different models.

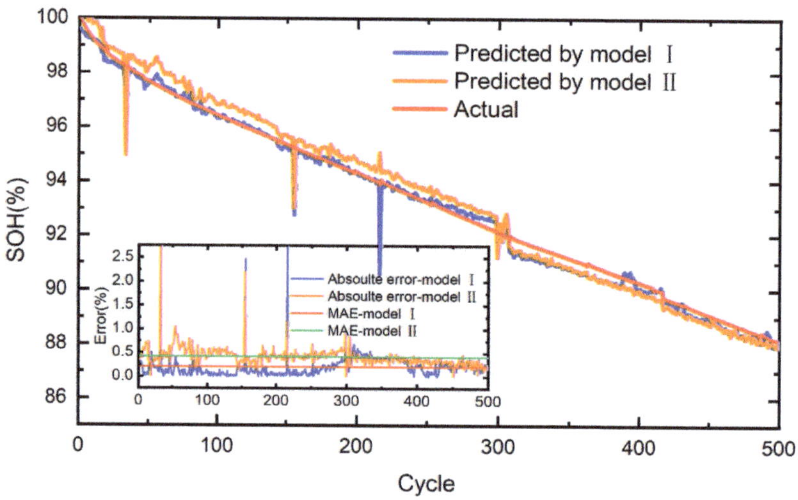

Figure 14. Results of Bat. #5 predicted by the trained model.

5. Conclusions

In this paper, we presented an innovative approach for estimating the SOH of Li-ion batteries by integrating incremental capacity analysis and internal resistance identification, using a hybrid model of traditional machine learning models and a KAN. The results demonstrate that the proposed hybrid model has good performance for SOH estimation in the test and validation dataset. The hybrid model integrated with the KAN effectively captured the complex relationships among the input features, leveraging its architecture to improve estimation performance. The accuracy of the hybrid models is higher than that of conventional models, which illustrates that hybrid models combined with KANs can enhance the accuracy and reliability of SOH predictions. The comparison of models with different inputs showed that including internal resistance as input parameters can enhance the accuracy of the model compared to using only IC features. All input features of the model were extracted from the charge data within the SOC range of 20% to 80%. The proposed model does not require full charge and discharge data. This framework shows promising application prospects in battery management systems of electric vehicles, which can facilitate more accurate monitoring and fast detection of battery SOHs.

Author Contributions: Writing—original draft, J.P.; funding acquisition, X.Z.; supervision, J.M.; investigation, D.M.; Methodology, S.J.; Formal analysis, K.Z.; visualization, C.G. and W.D. All authors have read and agreed to the published version of the manuscript.

Funding: This work was partially supported by the Shaanxi Postdoctoral Science Foundation (2023BSHEDZZ222), Fundamental Research Funds for the Central Universities, CHD (300102223103), National Natural Science Foundation of China (52172362, 52372375), Scientific and technological achievements transformation plan project of Shaanxi Province (2024CG-CGZH-19), and Key research and development plan project of Shaanxi Province (2024GX-YBXM-260).

Data Availability Statement: The original contributions presented in the study are included in the article, further inquiries can be directed to the corresponding author.

Conflicts of Interest: The authors declare no conflicts of interest.

References

1. Peng, J.; Zhou, X.; Jia, S.; Jin, Y.; Xu, S.; Chen, J. High precision strain monitoring for lithium ion batteries based on fiber Bragg grating sensors. *J. Power Sources* **2019**, *433*, 226692. [CrossRef]
2. Lu, L.; Han, X.; Li, J.; Hua, J.; Ouyang, M. A review on the key issues for lithium-ion battery management in electric vehicles. *J. Power Sources* **2013**, *226*, 272–288. [CrossRef]
3. Berecibar, M.; Gandiaga, I.; Villarreal, I.; Omar, N.; Van Mierlo, J.; Van den Bossche, P. Critical review of state of health estimation methods of Li-ion batteries for real applications. *Renew. Sustain. Energy. Rev.* **2016**, *56*, 572–587. [CrossRef]
4. Ng, K.S.; Moo, C.-S.; Chen, Y.-P.; Hsieh, Y.-C. Enhanced coulomb counting method for estimating state-of-charge and state-of-health of lithium-ion batteries. *Appl. Energy* **2009**, *86*, 1506–1511. [CrossRef]
5. Lu, J.; Xiong, R.; Tian, J.; Wang, C.; Sun, F. Deep learning to estimate lithium-ion battery state of health without additional degradation experiments. *Nat. Commun.* **2023**, *14*, 2760. [CrossRef]
6. Song, S.; Fei, C.; Xia, H. Lithium-Ion Battery SOH Estimation Based on XGBoost Algorithm with Accuracy Correction. *Energies* **2020**, *13*, 812. [CrossRef]
7. Chen, L.; Lü, Z.; Lin, W.; Li, J.; Pan, H. A new state-of-health estimation method for lithium-ion batteries through the intrinsic relationship between ohmic internal resistance and capacity. *Measurement* **2018**, *116*, 586–595. [CrossRef]
8. Li, X.; Wang, Z.; Zhang, L.; Zou, C.; Dorrell, D.D. State-of-health estimation for Li-ion batteries by combing the incremental capacity analysis method with grey relational analysis. *J. Power Sources* **2019**, *410–411*, 106–114. [CrossRef]
9. Weng, C.; Cui, Y.; Sun, J.; Peng, H. On-board state of health monitoring of lithium-ion batteries using incremental capacity analysis with support vector regression. *J. Power Sources* **2013**, *235*, 36–44. [CrossRef]
10. Zhang, M.; Chen, W.; Yin, J.; Feng, T. Lithium Battery Health Factor Extraction Based on Improved Douglas–Peucker Algorithm and SOH Prediction Based on XGboost. *Energies* **2022**, *15*, 5981. [CrossRef]
11. Wen, J.; Chen, X.; Li, X.; Li, Y. SOH prediction of lithium battery based on IC curve feature and BP neural network. *Energy* **2022**, *261*, 125234. [CrossRef]
12. Fan, Z.; Zi-xuan, X.; Ming-hu, W. State of health estimation for Li-ion battery using characteristic voltage intervals and genetic algorithm optimized back propagation neural network. *J. Energy Storage* **2023**, *57*, 106277. [CrossRef]
13. Li, W.; Sengupta, N.; Dechent, P.; Howey, D.; Annaswamy, A.; Sauer, D.U. Online capacity estimation of lithium-ion batteries with deep long short-term memory networks. *J. Power Sources* **2021**, *482*, 228863. [CrossRef]
14. Gong, Y.; Zhang, X.; Gao, D.; Li, H.; Yan, L.; Peng, J.; Huang, Z. State-of-health estimation of lithium-ion batteries based on improved long short-term memory algorithm. *J. Energy Storage* **2022**, *53*, 105046. [CrossRef]
15. Liu, Z.; Wang, Y.; Vaidya, S.; Ruehle, F.; Halverson, J.; Soljačić, M.; Hou, T.Y.; Tegmark, M. Kan: Kolmogorov-arnold networks. *arXiv* **2024**, arXiv:2404.19756.
16. Vaca-Rubio, C.J.; Blanco, L.; Pereira, R.; Caus, M. Kolmogorov-arnold networks (kans) for time series analysis. *arXiv* **2024**, arXiv:2405.08790.

17. Pastor-Fernández, C.; Uddin, K.; Chouchelamane, G.H.; Widanage, W.D.; Marco, J. A Comparison between Electrochemical Impedance Spectroscopy and Incremental Capacity-Differential Voltage as Li-ion Diagnostic Techniques to Identify and Quantify the Effects of Degradation Modes within Battery Management Systems. *J. Power Sources* **2017**, *360*, 301–318. [CrossRef]
18. Zheng, L.; Zhu, J.; Lu, D.D.-C.; Wang, G.; He, T. Incremental capacity analysis and differential voltage analysis based state of charge and capacity estimation for lithium-ion batteries. *Energy* **2018**, *150*, 759–769. [CrossRef]
19. Jiang, B.; Dai, H.; Wei, X. Incremental capacity analysis based adaptive capacity estimation for lithium-ion battery considering charging condition. *Appl. Energy* **2020**, *269*, 115074. [CrossRef]
20. Li, Y.; Abdel-Monem, M.; Gopalakrishnan, R.; Berecibar, M.; Nanini-Maury, E.; Omar, N.; van den Bossche, P.; Van Mierlo, J. A quick on-line state of health estimation method for Li-ion battery with incremental capacity curves processed by Gaussian filter. *J. Power Sources* **2018**, *373*, 40–53. [CrossRef]
21. Li, X.; Yuan, C.; Wang, Z. State of health estimation for Li-ion battery via partial incremental capacity analysis based on support vector regression. *Energy* **2020**, *203*, 117852. [CrossRef]
22. Beltran, H.; Sansano, E.; Pecht, M. Machine learning techniques suitability to estimate the retained capacity in lithium-ion batteries from partial charge/discharge curves. *J. Energy Storage* **2023**, *59*, 106346. [CrossRef]
23. Raccichini, R.; Amores, M.; Hinds, G. Critical Review of the Use of Reference Electrodes in Li-Ion Batteries: A Diagnostic Perspective. *Batteries* **2019**, *5*, 12. [CrossRef]
24. Barai, A.; Uddin, K.; Widanage, W.D.; McGordon, A.; Jennings, P. A study of the influence of measurement timescale on internal resistance characterisation methodologies for lithium-ion cells. *Sci. Rep.* **2018**, *8*, 21. [CrossRef] [PubMed]
25. Meddings, N.; Heinrich, M.; Overney, F.; Lee, J.-S.; Ruiz, V.; Napolitano, E.; Seitz, S.; Hinds, G.; Raccichini, R.; Gaberšček, M.; et al. Application of electrochemical impedance spectroscopy to commercial Li-ion cells: A review. *J. Power Sources* **2020**, *480*, 228742. [CrossRef]
26. Tan, X.; Tan, Y.; Zhan, D.; Yu, Z.; Fan, Y.; Qiu, J.; Li, J. Real-Time State-of-Health Estimation of Lithium-Ion Batteries Based on the Equivalent Internal Resistance. *IEEE Access* **2020**, *8*, 56811–56822. [CrossRef]
27. Galeotti, M.; Cinà, L.; Giammanco, C.; Cordiner, S.; Di Carlo, A. Performance analysis and SOH (state of health) evaluation of lithium polymer batteries through electrochemical impedance spectroscopy. *Energy* **2015**, *89*, 678–686. [CrossRef]
28. Grandjean, T.; Groenewald, J.; McGordon, A.; Widanage, W.; Marco, J. Accelerated Internal Resistance Measurements of Lithium-Ion Cells to Support Future End-of-Life Strategies for Electric Vehicles. *Batteries* **2018**, *4*, 49. [CrossRef]
29. Mei, W.; Jiang, L.; Liang, C.; Sun, J.; Wang, Q. Understanding of Li-plating on graphite electrode: Detection, quantification and mechanism revelation. *Energy Storage Mater.* **2021**, *41*, 209–221. [CrossRef]
30. Chahbaz, A.; Meishner, F.; Li, W.; Ünlübayir, C.; Uwe Sauer, D. Non-invasive identification of calendar and cyclic ageing mechanisms for lithium-titanate-oxide batteries. *Energy Storage Mater.* **2021**, *42*, 794–805. [CrossRef]
31. Ren, B.; Xie, C.; Sun, X.; Zhang, Q.; Yan, D. Parameter identification of a lithium-ion battery based on the improved recursive least square algorithm. *IET Power Electron.* **2020**, *13*, 2531–2537. [CrossRef]

batteries

MDPI

Article

General Machine Learning Approaches for Lithium-Ion Battery Capacity Fade Compared to Empirical Models

Quentin Mayemba [1,2,3,*], Gabriel Ducret [4], An Li [1], Rémy Mingant [2] and Pascal Venet [3]

1 Siemens Digital Industries Software, 19 Boulevard Jules Carteret, 69007 Lyon, France; an.li@siemens.com
2 IFP Energies Nouvelles, Rond-Point de L'échangeur de Solaize, 69360 Solaize, France; remy.mingant@ifpen.fr
3 Université Claude Bernard Lyon 1, Ampère, UMR5005, INSA Lyon, Ecole Centrale de Lyon, CNRS, F-69100 Villeurbanne, France
4 IFP Energies Nouvelles, 1-4 Avenue du Bois Préau, 92500 Rueil-Malmaison, France; gabriel.ducret@ifpen.fr
* Correspondence: quentin.mayemba@siemens.com

Abstract: Today's growing demand for lithium-ion batteries across various industrial sectors has introduced a new concern: battery aging. This issue necessitates the development of tools and models that can accurately predict battery aging. This study proposes a general framework for constructing battery aging models using machine learning techniques and compares these models with two existing empirical models, including a commercial one. To build the models, the databases produced by EVERLASTING and Bills et al. were utilized. The aim is to create universally applicable models that can address any battery-aging scenario. In this study, three types of models were developed: a vanilla neural network, a neural network inspired by extreme learning machines, and an encoder coupled with a neural network. The inputs for these models are derived from established knowledge in battery science, allowing the models to capture aging effects across different use cases. The models were trained on cells subjected to specific aging conditions and they were tested on other cells from the same database that experienced different aging conditions. The results obtained during the test for the vanilla neural network showed an RMSE of 1.3% on the Bills et al. test data and an RMSE of 2.7% on the EVERLASTING data, demonstrating similar or superior performance compared to the empirical models and proving the ability of the models to capture battery aging.

Keywords: capacity loss; battery aging; empirical model; machine learning; artificial neural network; autoencoder

1. Introduction

The automotive industry is transitioning toward electric mobility. Fossil fuels are not suitable for storing energy in electric vehicles, which is why lithium-ion batteries are, nowadays, the common preference. In addition, this trend has been emphasized by the need to store the intermittent electricity produced by renewables and by an increase in the number of portable electronic devices being sold. These factors contribute to a growing market in the field of energy storage. In this context, lithium-ion batteries emerge as a viable solution [1]. They are energetically dense and financially affordable, and they also have a relatively high cycle life and efficiency.

In spite of their relatively high cycle life, numerous degradation phenomena can occur in these accumulators, as shown in Table 1, adapted from Ref. [2]. The working principle of a fresh cell, along with several aging phenomena, is illustrated in Figure 1 (the crystal structure of the LCO active material has been represented using chemtube3d [3]). Several factors significantly influence the aging of the cells, including the current, temperature, state of charge (SOC), voltage, and mechanical stress. These factors are referred to as stress factors. In the models developed for this study, only the current, temperature, and SOC are considered.

Table 1. Lithium-ion accumulator's degradation phenomena, adapted from Ref. [2].

Number	Phenomenon	Reference
1	Solid Electrolyte Interphase (SEI) growth, which can be considered the most important degradation phenomenon (the composition and behavior are different in silicon-containing negative electrodes)	[4,5]
2	Lithium plating and dendrite growth	[6,7]
3	Particle cracking, which is due to the volume changes during lithiation	[7–9]
4	Gas bubbles formation and electrolyte drying (mainly the formation of H_2, due to the decomposition of the organic molecules in the electrolyte)	[7,10,11]
5	Structure changes in the active material happen when the crystal structure of the active material changes and lithium cannot be inserted any more	[12]
6	Transition metals dissolution (the transition metals concerned are mainly Ni, Co, and Mn, dissolving into the electrolyte)	[8]
7	Graphite exfoliation and solvent co-intercalation, happens when the electrolyte solvent inserts itself into the graphite with the lithium-ion and separates the graphite sheets	[13,14]
8	The growth of the positive electrode–electrolyte interface	[15]
9	The corrosion or dissolution of the current collectors	[16]
10	The loss of electric contact	[12,17,18]
11	The decomposition of the binders	[19]
12	The decomposition of the electrolyte	[12]
13	The degradation of the separator	[20]

This degradation results in a loss of capacity and an increase in the internal resistance of the cell. The capacity corresponds to the electric charge (expressed in Amp-hours) that can be stored in the battery. It is correlated to the amount of lithium that can pass from one electrode to the other. Predicting the aging process of lithium-ion cells is crucial because it allows manufacturers to ensure that their lithium-ion cells can consistently meet the product's energy specifications over time. Lithium-ion battery aging forecasting is also key to assessing batteries' potential for reuse, in order to give them a second life [21] or to assess their safety [22]. The development of battery degradation models, therefore, has become necessary. The literature categorizes these models into three main categories:

- physical models [23,24];
- empirical models [25,26];
- machine learning (ML) models [27,28].

It should, however, be emphasized that some models rely on several of these categories, such as grey-box models that combine physics and data-driven methods, as discussed in the review by Guo et al. in 2022 [29]. Physical models have the advantage of being highly explainable, but they also have drawbacks. They require a significant amount of experimental data, often obtained by dismantling the battery cell, and their application necessitates a high computational cost. This cost can be explained by the resolution of numerous partial differential equations in electrochemical models that consider aging phenomena. Examples of such models are the single particle model, the single particle model with electrolyte dynamics, and the pseudo-2 dimensional model [24]. It is important to note that the current state-of-the-art physical models do not encompass all the aging phenomena described in Table 1. Instead, they focus on a select few phenomena [10,23,24], such as:

- SEI growth;
- lithium-plating;
- particle cracking due to mechanical stress (and the associated growth of the SEI on the cracks);
- loss of active material (LAM) due to mechanical stress and the internal cracks of the particles;
- oxidation of the electrolyte at the positive electrode.

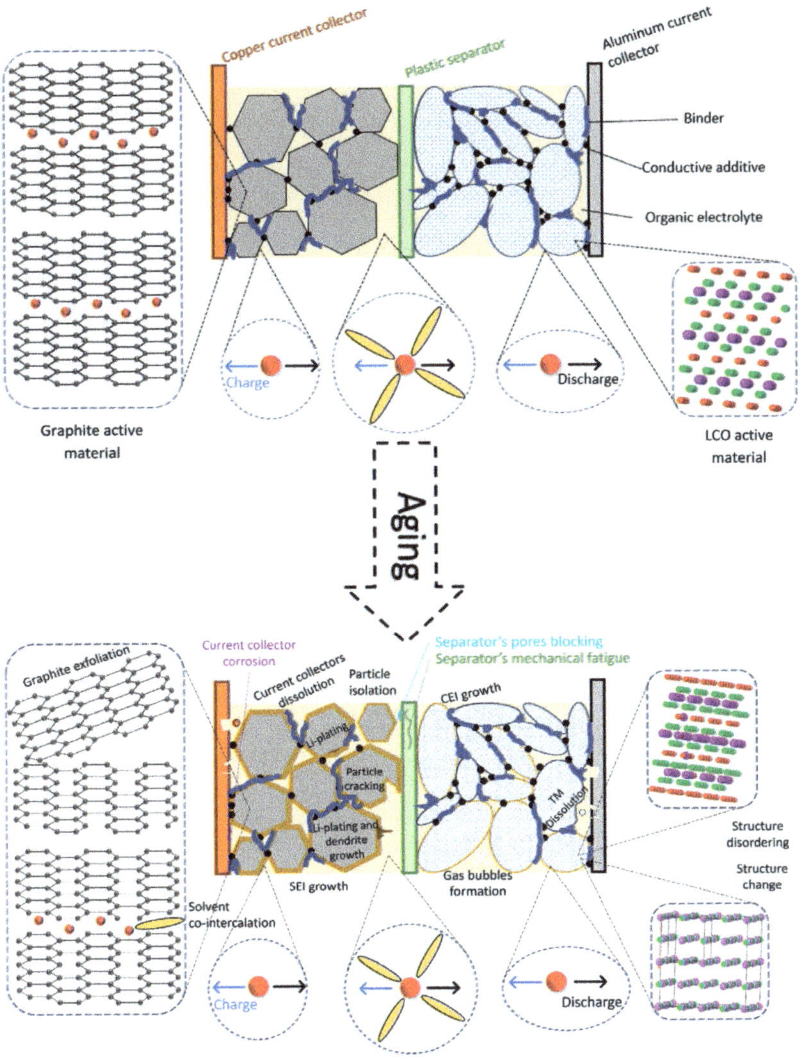

Figure 1. Schematic representation of some aging phenomena occurring in a lithium-ion cell. In the figure, the red disks correspond to lithium atoms and lithium ions, the grey disks correspond to carbon atoms, the purple disks correspond to cobalt atoms, and the green disks correspond to oxygen atoms.

The two other categories of aging models are data-driven, meaning that they rely on battery-aging data and more specifically on capacity fade measurements. These models are consequently able to capture any aging phenomenon occurring in the cell but are unable to attribute the degradation to specific phenomena. The distinction between empirical and machine learning models is that the ML models utilize the tools developed by the machine learning community to create agnostic models. These ML models do not assume any pre-existing knowledge, nor do they make specific assumptions about the data. In contrast, empirical models rely on mathematical formulas created by the designer of the model and are based on the available data. ML models often have more parameters to optimize than empirical models. The scope of this study is limited to models that predict the aging of cells regardless of the specific aging phenomena occurring within them. This study consequently focuses on data-driven models, and more specifically those implementable in simulation software to estimate the aging of lithium-ion cells under various aging conditions. Only battery aging models based on non-linear regression with exogenous variables are considered, which means that previous outputs are not utilized as inputs. Furthermore, the models discussed in this paper do not depend on the response of the cell to a specific stimulation. Thus, these models may rely on the current that a cell delivers during aging but not on its electrothermal response to a specific test conducted during a check-up, for instance. Part of the incremental capacity curves or any other health indicators (presented in Section 4.3.1 of the review by Khaleghi et al. in 2024 [30]) cannot, therefore, be used as inputs of the models considered. That is why models that rely on previous capacity measurements are also excluded [31]. In this study, neural networks are employed to forecast the capacity loss of the cells, based on the time they spent under specific conditions. New features have been proposed and applied to databases where such methods had never been previously implemented, thus distinguishing our approach from the state of the art. The effectiveness of this method was assessed by the use of two diverse databases that encompass a wide range of applications, including driving profiles, calendar aging, constant current cycling, and electric vertical take-off and landing (eVTOL) aircraft operations. The results thus obtained were then compared with those from two empirical models. To the extent of the authors' knowledge, this is the first instance in which this method is compared to empirical models and applied to open-source data covering various ambient temperatures and current profiles.

This paper is organized as follows: Section 2 provides an overview of state-of-the-art machine-learning models that forecast the capacity loss of lithium-ion batteries, while Section 3 describes the databases, the empirical models developed, and the machine-learning models used here. Finally, Section 4 presents the results obtained from each database using the methods mentioned.

2. State of the Art on Machine Learning Models to Predict Battery Aging Using Time Passed by the Cell between Thresholds

Building on an understanding of the challenges related to lithium-ion batteries, approaches that harness machine learning techniques are discussed below. These methods predict battery degradation accurately by utilizing extensive datasets to effectively model complex aging dynamics [32]. The literature on the application of machine learning to study the state of health of lithium-ion cells is extensive, with numerous reviews having been published in recent years [27,28,30,33]. While listing all the methodologies would be tedious due to their vast diversity, the review of von Bülow et al. in 2023 [27] highlights that fewer papers focus on developing models that can predict the SOH of cells as tools for engineers. Such a tool could help identify which types of real-world usage most significantly impact a battery's health. The present study specifically targets such models. The most versatile models employ historical features, sometimes also referred to as histograms. The fundamental concept behind these models is to make categories, or zones, in the cells' stress factors (such as the temperature, current, voltage, SOC, or mechanical stress). As inputs to the ML model, these zones consider how long the cell has remained in each category, either through the duration spent in the zone or the count of measurements taken while the cell was in that zone. To the best of the authors' knowledge, no other information has been extracted from these zones to be used as input into machine learning models. As discussed in Section 3.3 of the present study, additional features have been incorporated, such as the time integral of temperature within the zone, the time integral of the current, and the time integral of the SOC. This information allows the model to evaluate the potential damage associated with the duration for which the cell remained in each zone.

The first authors to propose the methodology of using thresholds to create categories and utilize information from these categories to predict a cell's aging were Nuhic et al. in 2013 [34]. Their objective was to develop battery aging models that were not specific to any particular discharge pattern. To achieve this, they studied what they referred to as the time the cell remained within specific parameter threshold values. They established 9 classes of SOC, 7 classes of temperature, and 13 classes of current. The authors then utilized the time spent by the cell in each combination of SOC and temperature classes (resulting in 7 times 9 inputs) and in each combination of current and temperature classes (resulting in 7 times 13 inputs) as the inputs for their model. In addition, they examined the rainflow counting of the SOC to obtain inputs that could represent the number of cycles without explicitly defining a cycle in the aging profile. These inputs were used to build a support vector regression system. They not only predicted the cell capacity from the beginning until each check-up but also between each check-up.

In 2016, You et al. [35] also conducted a similar study in which they utilized the current, voltage, and temperature measurements of cells. They applied the K-means clustering algorithm to partition the resulting three-dimensional space into 80 distinct regions. Then, they employed the densities of each of these 80 clusters as inputs for support vector machines (SVMs) and an artificial neural network (ANN) to predict the SOH of the cells. To create their dataset, they subjected the cells to cycling profiles inspired by the various driving conditions encountered with electric vehicles, such as highway and urban driving. This approach is comprehensive as it allows for the consideration of various aging profiles. However, it is important to note that analyzing the density of the regions may introduce bias if the sampling frequency varies throughout the cells' lifespan. For instance, measurements may be gathered with a larger timestep during constant current charging, compared to during driving cycles.

Considering functions of time rather than the number of measurements appears to be more reliable. This idea was proposed by Richardson et al. in 2019 [36], although they did not use such features to construct their Gaussian process regression (GPR) model using the NASA randomized battery dataset. In their study, they suggested considering the time spent between the two thresholds of a specific parameter, such as current, temperature, voltage, or power. Consequently, their approach can be referred to as one-dimensional histograms, or 1D models, in contrast to the three-dimensional models made by You et al. Their GPR model used the Ah-throughput of the cells, their age, and the time between two consecutive capacity measurements to predict the corresponding capacity change, expressed in Ah. It is noteworthy that such approaches require knowledge of the distribution of the stress factors; this is why they are often referred to as histograms.

In 2020, Song et al. [37] used a similar method. They utilized real-world data from 300 battery-powered electric vehicles and 400 hybrid vehicles, collected between January 2018 and December 2018, to predict the capacity of these batteries. The variables considered included the mileage of the vehicles and the distribution of measurement points across several thresholds for current, SOC, and temperature. These inputs were then transformed using principal component analysis [38] and the resulting components were ordered according to the variance they explained in the dataset. Only those components with a cumulative variance below 90% were kept as inputs of the neural network. Given that the vehicles were being used by the customers, the cells experienced no capacity measurement per se. The capacities were estimated during charging at between 40% and 80% SOC.

In 2021, von Bülow et al. [39] compared two-dimensional and three-dimensional models using the Severson et al. database [40] to predict SOH changes for a given number of cycles. The three stress factors that they focused on were current, temperature, and SOC. Their two-dimensional models relied on the time spent by the cell between the thresholds of current and temperature, then those on current and SOC, and finally those on SOC and temperature. One model considered all these features as inputs. They studied the influence of window width (or the number of cycles for which ΔSOH was predicted) and different thresholds of the stress factors to retrieve their inputs, then they built an ANN and a GPR. The hyperparameters of the ANN were optimized using a tree-structured Parzen estimator (TPE). The hyperparameters in a machine learning model are the characteristics that define the model, such as the number of layers in an ANN or the number of neurons per layer [41]. They found that combining multiple window widths during training (i.e., training the model to predict capacity loss for not only 50 equivalent full cycles ahead but also 25 or 100 equivalent full cycles ahead) enhanced its ability to generalize. In their experiment, the 3D models did not yield better results than the 2D ones. However, it is worth noting that the dataset they employed did not include any variation in the temperature of the thermal chamber. Therefore, temperature differences between the cells were attributed solely to their electro-thermal behavior.

In 2022, Zhang et al. [42] utilized historical features from three datasets: the study by Severson et al. and Attia et al. [40,43], the NASA randomized battery dataset [44], and a dataset based on actual measurements from plug-in hybrid electric vehicles (PHEV). The models employed included random forest regressions, support vector regressions, ANN, and GPR. The authors used the historical features of the cells to forecast their degradation trajectories.

In 2023, Greenbank et al. [45] conducted a histogram-based approach to determine the thresholds for voltage, temperature, and current, based on the 1st, 33rd, 67th, and 99th percentiles of these stress factors. However, the temperature-related features were excluded from the model because they did not pass the feature selection process. They utilized datasets from Severson et al. [40] and Attia et al. [43] to predict the knee-point and remaining useful life of the cells. Specifically, they estimated the cells' capacity loss every 12 h, which roughly corresponds to 9 to 19 cycles, depending on the charging profiles. The knee point is defined as the moment when the slope of aging over time significantly increases [46].

The contents of these works, along with what is proposed here, are summarized in Table 2.

Table 2. Summary of previous, similar approaches.

Study	Dataset	Open-Source Data?	Method to Fix Thresholds	Inputs	Outputs	Model	Temperature
Nuhic 2013 [34]	5 batteries, no reference to another work or the data publication	No	Not shared	2D inputs (I&T and SOC&T) and rainflow counting of the SOC	ΔQ	SVR	Yes
You 2016 [35]	Private	No	K-Means clustering	Density of the 3D zones of (I, T, SOC)	SOH	SVM, ANN	Yes
Richardson 2019 [36]	NASA Randomized Battery Dataset	Yes	Arbitrary (Proposed 1D data but did not implement it)	Ah-throughput, Δt, and time.	ΔQ	GPR	Proposed to consider it but did not implement
Song 2020 [37]	BHEV and BEV data from real vehicles	No	Arbitrary	Principal components of the mileage and the 1D distribution of current, SOC, and temperature	Q	ANN	Yes

Table 2. *Cont.*

Study	Dataset	Open-Source Data?	Method to Fix Thresholds	Inputs	Outputs	Model	Temperature
von Bülow 2021 [39]	Severson [40]	Yes	Arbitrary	2D, 3D	ΔSOH	ANN and GPR	Yes
Zhang 2022 [42]	Severson [40] Attia [43], NASA randomized battery dataset [44], and a dataset based on actual measurements from plug-in hybrid electric vehicles (PHEV).	Partly	Not shared	Derived from the histogram data	$\Delta Q_{k\ cycles\ ahead}$	Random forest regressions, support vector regressions, ANN, and GPR	Yes
Greenbank 2023 [45]	Severson [40] Attia [43]	Yes	Based on the distribution	Time spent (1D)	$Q_{loss\ t+12\ h}$	Piecewise linear model, Gaussian process regression	Yes
Present work	EVERLASTING [47] Bills [48]	Yes	Arbitrary	Time spent (1D) and time integrals (3D)	Q_{loss}	ANN, ELM, auto-encoders	Yes

In conclusion, machine learning models provide a robust framework for the predictive analysis of lithium-ion battery aging. These models utilize historical data to generate accurate forecasts. Building upon the literature, three models based on the time spent in various zones are presented in Section 3.3. These models have been compared with the empirical models discussed in Section 3.2. Both modeling approaches have been applied to the data presented in Section 3.1, which are open-source data wherein cells were aged under different cycling procedures and storage temperatures. The method relies on the time spent by the cell between temperature thresholds, among other parameters. The following points make these models original:

- The thresholds selected in previous studies were specific to the datasets used. In contrast, the present work proposes general thresholds applicable to any battery aging dataset, thus contributing to moving a step closer to a general battery aging model.
- Except for the study by Zhang et al. [42], which employed the NASA randomized battery dataset [44], the models available in the literature have not been applied to open-source databases where all cells do not experience the same ambient temperature during aging. Consequently, this paper presents the results from applying the method to two additional public datasets, wherein temperature thresholds are associated not only with cell heating but also with varying aging conditions, resulting in sparser data.
- This paper demonstrates the robustness of the method when applied to public datasets, thereby advancing the development of generalized battery aging models. These machine learning methods have not been previously applied to these datasets, and their usage introduces challenges that are distinct from those encountered before.
- Furthermore, to the best of the authors' knowledge, this study is the first application of autoencoders for reducing the dimensionality of inputs based on the time spent.
- This study introduces novel inputs derived from zones that are not based on the time spent in a zone or the zone's density. Instead, these novel inputs include the time

integral of the current in the zone, that of the SOC, and that of the temperature. This approach enables the model to account for parameter variations within a zone.

3. Materials and Methods

3.1. Battery Aging Datasets Used

In the present study, the models were applied to two databases: EVERLASTING [47] and that of Bills et al. from 2023 [48]. Although additional battery aging databases exist, such as those referenced in Refs. [2,49,50], the focus of this study is on these two databases because they encompass a range of scenarios inspired by real-world usage cases. The description of these datasets can be found in Ref. [2].

3.1.1. The EVERLASTING Dataset

The EVERLASTING dataset project studied the commercial cells named INR18650 MJ1, manufactured by LG Chem, with a nominal capacity of 3.5 Ah. These cells incorporate positive electrode active material composed of NMC 811. The active material of the negative electrode is made of a blend of graphite intercalation compound (GIC) and SiO_x. The test matrix conducted in this project is summarized in Tables 3–5. The resulting capacity losses for 40 cells are illustrated in Figure 2. In this figure, it can be seen that, generally speaking, calendar aging at 0 °C and 10 °C produced similar results, with less degradation being observed compared to calendar aging at 25 °C. The most severe calendar aging occurred at 45 °C. Across all temperatures, capacity retention was superior during calendar aging compared to CC cycling. The driving profiles, along with CC cycling at 10 °C, exhibited the most heterogeneous characteristics regarding cell degradation.

Table 3. Number of cells for life cycle conditions using driving profiles in EVERLASTING, sourced from Ref. [2].

Temperature	I Profile 70–90% SOC	I Profile 10–90% SOC	P Profile 10–90% SOC
0 °C	2	2	
10 °C	2	2	
25 °C	2	2	2
45 °C	2	2	

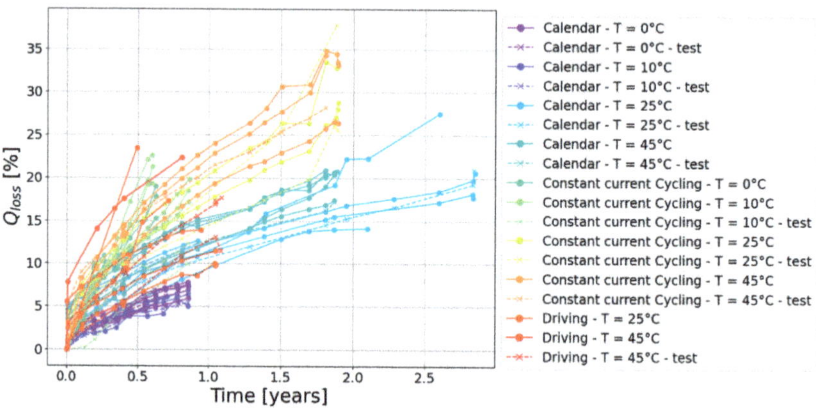

Figure 2. Capacity loss versus time in the EVERLASTING database. Half of the test conditions were never seen in the training data (considering the SOC, charge, and discharge currents not represented here). The other half correspond to conditions that can be found in the training data but were experienced by cells that were not in the training data.

Table 4. Constant DC cycle aging conditions in EVERLASTING, sourced from Ref. [2].

Temperature	Charge C-Rate	Discharge C-Rate	Number of Cells
0 °C	0.5	1.5	2
0 °C	1	1.5	2
10 °C	0.5	1.5	2
10 °C	0.5	0.5	2
10 °C	0.5	3	2
10 °C	1	1.5	2
25 °C	0.5	1.5	2
25 °C	0.5	0.5	2
25 °C	0.5	3	2
25 °C	1	1.5	2
45 °C	0.5	1.5	2
45 °C	0.5	0.5	2
45 °C	0.5	3 (a)	2
45 °C	1	1.5	2

(a): These cells could not cycle because the temperature (45 °C) and the discharge C-rate (3C) caused the cells to heat up to 55 °C too rapidly. Only one capacity measurement was available under these conditions.

Table 5. Number of cells under calendar aging conditions in EVERLASTING, sourced from Ref. [2].

SOC Temperature	10%	70%	90%
0 °C	2	2	2
10 °C	2	2	2
25 °C	2	2	2
45 °C	2	2	2

All the cells were included in the training set, except those in the test set described below. In this database, the test data comprised cells 19, 20, 34, 37, 63, 68, 69, 72, 78, 79, and 96. These cells are represented in Figure 2 by dashed lines marked with crosses. The aging conditions for these cells were as follows:

- Cells 19 and 20 experienced CC cycle aging at 10 °C;
- Cell 34 underwent calendar aging at 10 °C and SOC = 90%;
- Cell 37 underwent calendar aging at 0 °C and SOC = 70%;
- Cell 63 experienced CC cycle aging at 45 °C;
- Cells 68 and 69 experienced driving aging at 45 °C between 70 and 90% SOC;
- Cell 72 underwent calendar aging at 45 °C and SOC = 10%;
- Cells 78 and 79 experienced CC cycle aging at 25 °C;
- Cell 96 underwent calendar aging at 25 °C and SOC = 90%.

The commercial model presented in Section 3.2 could not be calibrated using driving profiles. Therefore, for this model, the test data also included all driving measurements. It is important to note that the training and testing sets of the models were composed of different aging conditions.

3.1.2. The Bills Dataset

The database of Bills et al. [48] was inspired by the behavior of electric vertical take-off and landing aircraft (also referred to as eVTOLs) [51]. In this database, 22 NCA cells with a nominal capacity of 3 Ah were cycled. The negative electrode was made of pure graphite.

The baseline power and current profiles are illustrated in Figure 3, and each phase of the baseline profile is explained in detail in Table 6. The temperature for the baseline profile was 25 °C.

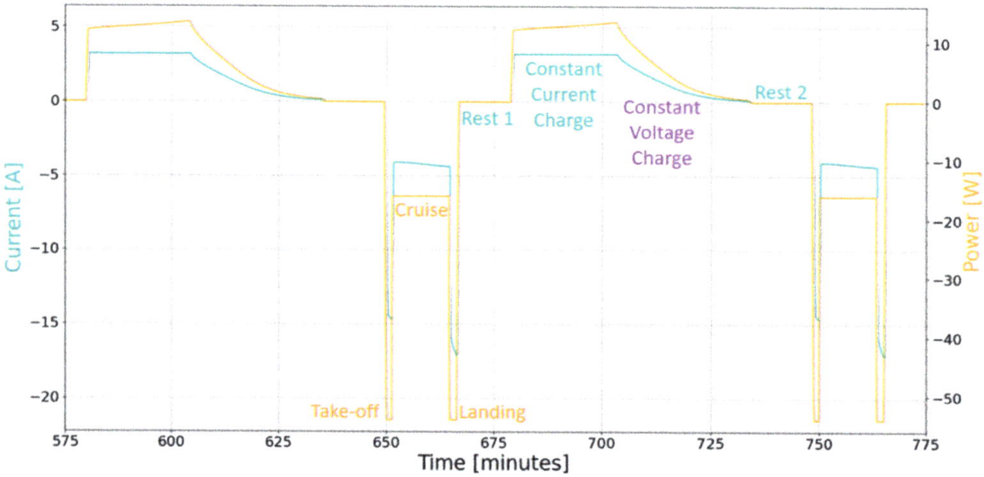

Figure 3. Baseline profile of the Bills et al. data. The graph was obtained from cell VAH27. The color corresponds to what is being imposed on the cell (i.e., the CC charge imposes the charge current, the CV charge imposes the charge voltage, and the discharge profile is imposed in power over a given time).

Table 6. Baseline mission parameters from Ref. [48], used under the license CC BY [52].

Phase	Definition	End Criteria
Take-off	P = 54 W	t = 75 s
Cruise	P = 16 W	t = 800 s
Landing	P = 54 W	t = 105 s
Rest 1	I = 0 A	T < 27 °C
CC Charge	I = 1 C	U > 4.2 V
CV Charge	U = 4.2 V	I < C/30
Rest 2	I = 0 A	T < 35 °C

The test matrix is presented in Table 7 and includes variations from the baseline scenario. The table specifies the number of data points, which is equal to the number of capacity measurements. The bold values represent the number of data points used as test data. Notably, only one test cell is aging in the conditions that were present in the training data.

Table 7. Aging profiles in the Bills et al. dataset [48].

Mission Profile	Cells Impacted	Number of Data Points
Baseline	VAH01, VAH17, VAH27	52
Short cruise (400 s)	VAH12	47
Short cruise (600 s)	VAH13, VAH26	45
Extended cruise (1000 s)	VAH02, VAH15, VAH22	**36**
10% power reduction during discharge	VAH05, VAH28	30 and **24**
20% power reduction during discharge	VAH11	44
CC charge current reduced to C/2	VAH06, VAH24	37
CC charge current brought up to 1.5 C	VAH16, VAH20	24
CV charge voltage reduced to 4.0 V	VAH07	6
CV charge voltage reduced to 4.1 V	VAH23	15
Thermal chamber temperature reduced to 20 °C	VAH09, VAH25	35
Thermal chamber temperature brought up to 30 °C	VAH10	**29**
Thermal chamber temperature brought up to 35 °C	VAH30	19

The resulting capacity loss versus time is illustrated in Figure 4. The tests continued until the cells either reached 70 °C or 2.5 V during discharge. As can be seen, aging is rather similar for all cells, regardless of aging parameters such as the room temperature or the maximal voltage of the cells. This can be explained by the very demanding discharge profile that causes all cells to heat up and reach temperatures above 40 °C.

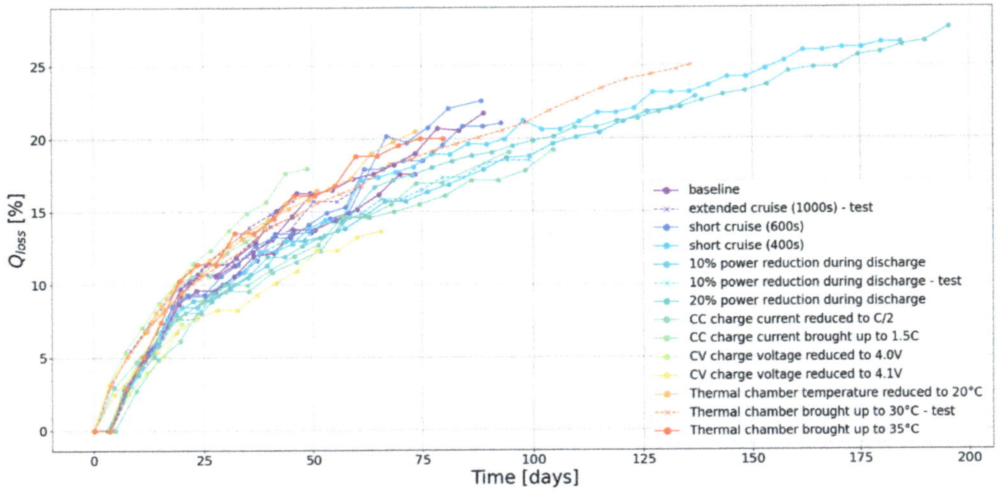

Figure 4. Capacity losses according to conditions in the Bills database.

In this dataset, the cells VAH02, VAH10, VAH15, VAH22, and VAH28 served as test data, while the remaining 17 cells were used as training data, as illustrated by the dashed line in Figure 4.

3.2. Empirical Models

This study employed two distinct empirical models to assess battery degradation:

1. a basic aging model;
2. a generic aging model, implementing the battery aging identification tool in the commercial software Simcenter Amesim V2310 [53]. This model is inspired by the work of Mingant et al., 2021 [25].

These two models were selected for comparison because one is a very simple model while the other is a state-of-the-art model implemented in commercial simulation software. The first one models degradation using an exponential function of time:

$$Q_{loss} = B \times t^{z} \tag{1}$$

where B and z are constants.

The commercial model is inspired by the following equation [25]:

$$\frac{dQ_{loss}}{dt} = B \times z \times \left(\frac{Q_{loss}}{B}\right)^{\frac{z-1}{z}} \tag{2}$$

where B and z follow the equations:

$$z = z_{min} + (z_{max} - z_{min}) \times \frac{1 + tanh\left(z_{poly}\right)}{2} \tag{3}$$

$$\begin{cases} B = B_{poly}\ if\ B_{poly} \geq 20 \\ B = 20 \times exp\left(\frac{B_{poly}-20}{20}\right)\ otherwise \end{cases} \tag{4}$$

with z_{poly} and B_{poly}:

$$z_{poly} = z_1 + z_2 \times T + z_3 \times SOC + z_4 \times I_{ch} + z_5 \times I_{dch} \tag{5}$$

$$\begin{aligned} B_{poly} = {} & B_1 + B_2 \times \tfrac{1}{T} + B_3 \times SOC + B_4 \times SOC \times T + B_5 \times T \times I_{ch} + B_6 \times \tfrac{I_{ch}}{T} + B_7 \times \tfrac{I_{dch}}{T} + B_8 \times \tfrac{SOC}{T} + B_9 \times T^2 \\ & + B_{10} \times I_{ch}^2 + B_{11} \times I_{dch}^2 + B_{12} \times SOC^2 \end{aligned} \tag{6}$$

where T corresponds to the temperature of the cell at time t, I_{dch} corresponds to the discharge current at the same time t, and I_{ch} corresponds to the charge current at the same time t. In this work, this model is referred to as the commercial model.

In this commercial model, the coefficients written in blue are optimized during the calibration process. The software is limited to optimizing these coefficients with a single charge current and a single discharge current. Therefore, the optimization was performed using the root mean square discharge current from the Bills database. However, the model was not calibrated on dynamic profiles, such as the driving profile; consequently, all driving profiles were utilized as test data. It is important to note that there are numerous other empirical aging models available [26,54], but only the aforementioned models were applied in this study.

The optimization of this model on the EVERLASTING data resulted in the following mathematical operations:

$$z_{poly} = 6.137 - 2.451 \times 10^{-2} \times T + 3.551 \times 10^{-1} \times I_{dch} + 8.777 \times 10^{-3} \times SOC \tag{7}$$

$$\begin{aligned} B_{poly} = {} & 627.5 - 1.345 \times 10^{5} \times \tfrac{1}{T} - 17.72 \times SOC + 2.926 \times 10^{-2} \times SOC \times T - 0.5205 \times T \times I_{ch} + 4.663 \times 10^{4} \times \tfrac{I_{ch}}{T} \\ & + 2.443 \times 10^{4} \times \tfrac{I_{dch}}{T} + 2.736 \times 10^{3} \times \tfrac{SOC}{T} - 1.926 \times 10^{-3} \times T^2 - 18.39 \times I_{ch}^2 - 40.20 \times I_{dch}^2 \\ & - 1.655 \times 10^{-3} \times SOC^2 \end{aligned} \tag{8}$$

This model has also been optimized on the Bills data, yielding the following results:

$$z_{poly} = 8.211 \times 10^{-2} - 8.530 \times 10^{-3} \times SOC \tag{9}$$

$$B_{poly} = 42.34 + 0.4137 \times T \times I_{ch} - 4.239 \times 10^{4} \times \frac{I_{ch}}{T} + 14.46 \times I_{ch}^2 \tag{10}$$

3.3. Machine Learning Models

Building upon the existing literature, models based on the time that the cells spent under specific conditions were created. To achieve this, thresholds for the three stress factors (i.e., state of charge, temperature, and current) considered in our models were defined. These thresholds were arbitrarily chosen but corresponded to battery science concepts, such as a cell with a low SOC or a high SOC, for instance. Importantly, these thresholds were consistent across the two databases studied. The model utilized the time the cell has spent between these thresholds to predict capacity loss. It was designed to perform effectively with various data sources, thereby ensuring consistent and reliable predictions. Three variants of this approach are considered here (see Figure 5).

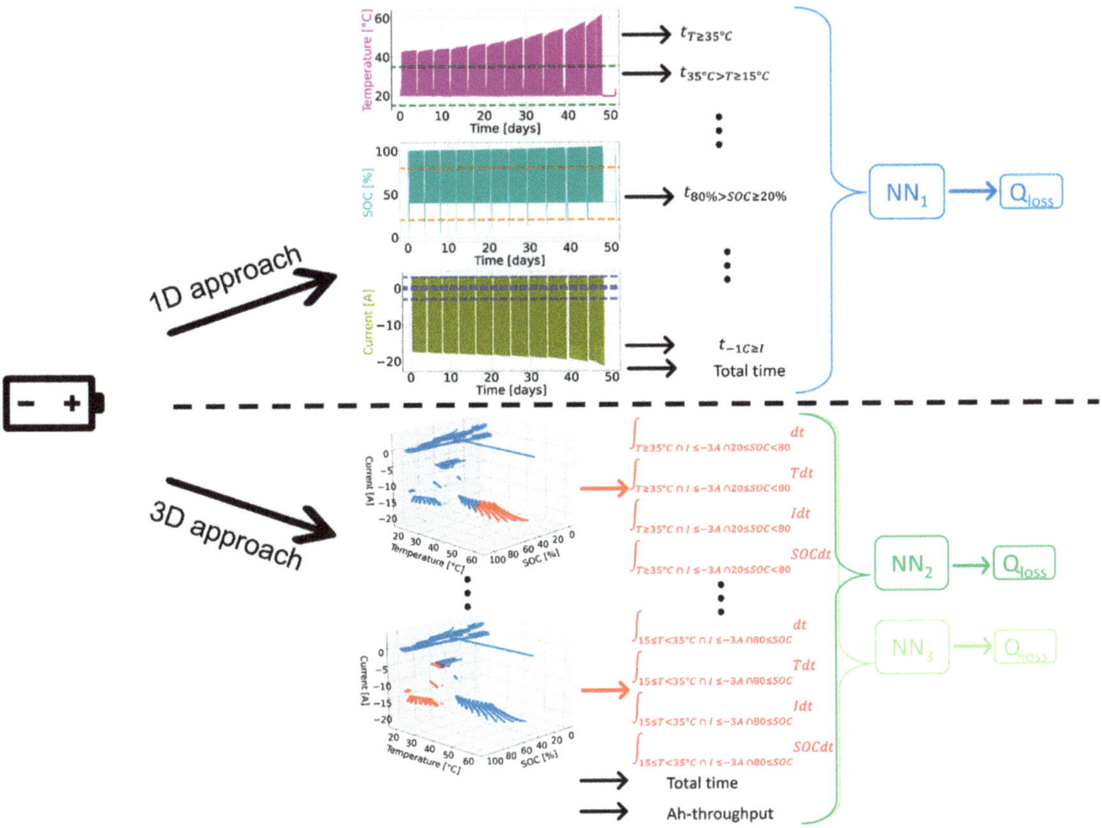

Figure 5. The two data transformation approaches and the three model architectures (neural networks). The graphs shown are for the VAH25 cell in the Bills dataset. The 3D plot represents all measurements taken on this cell, and those that satisfy the zone defined on the right are represented in red, whereas the others are in blue. The 3D models are represented in green and the 1D model is represented in blue.

The first neural network (NN_1), or 1D zone, is designed to determine the duration the cell has spent with one stress factor between two thresholds, independent of the values of the other two factors. For example, the time spent by the cell with a SOC of between 20% and 80% was used as input. The architecture of the neural networks built for neural network 1 is displayed in Figure 6.

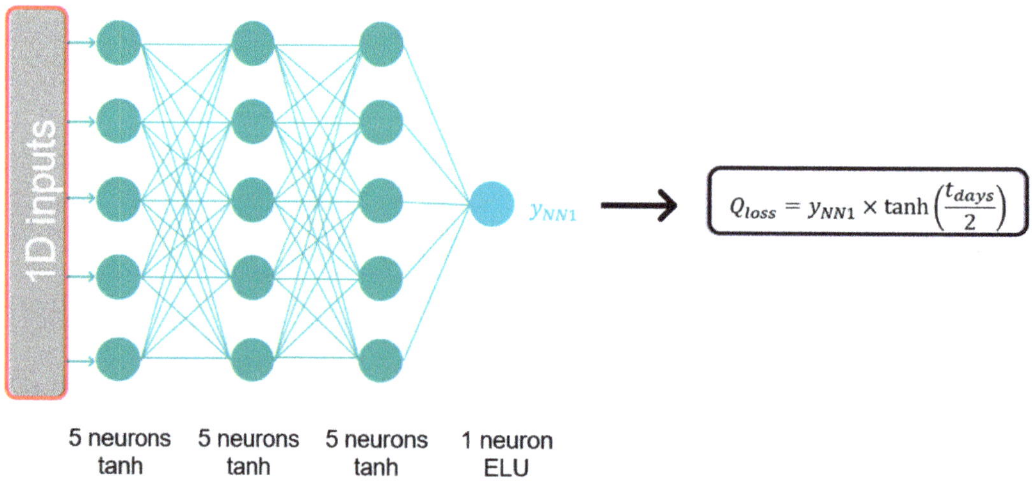

Figure 6. Neural network architecture 1 (NN_1): the model from the 1D data.

The other two neural networks adopt a 3D approach to data transformation. In this approach, the thresholds are applied simultaneously to every stress factor. This creates distinct zones within the space of stress factors. Within this space, the inputs selected for the neural networks include the time spent by the cell in these zones, along with the time integral of the current, the temperature, and the SOC in each zone. All these inputs are utilized by neural networks 2 and 3 to predict the capacity losses. Neural network 2 (NN_2) features an architecture similar to that of neural network 1, but it does not train its first layer, which reduces the number of parameters needing optimization. The architecture of NN_2 is illustrated in Figure 7. This neural network draws inspiration from an extreme learning machine (ELM) [55,56]. The decision to exclude training of the first layer is pertinent given the high dimensionality of the 3D input data. Neural network 3 (NN_3), however, incorporates a different mechanism: it performs data reduction through an encoder that is followed by a neural network. Autoencoders consist of two components: an encoder and a decoder. The encoder transforms the input into a lower dimensional space, while the decoder reconstructs the original input, based on the encoded representation. By using only the encoder part of the autoencoder, a dimensionality reduction technique is applied [57,58]. As shown in Figure 8, an autoencoder was constructed using the input data, and its encoding component was reutilized in the model to predict capacity loss. To develop the autoencoder, the data were augmented by multiplying the inputs by factors ranging from 0.1 to 3 in 0.1 increments, effectively increasing the number of data points by a factor of 29. This augmentation is feasible because the inputs pertain to the time spent by the cell under specific conditions. Thus, multiplying the inputs by certain factors simulates having a cell that ages under specific conditions for various durations. A possible physical interpretation of these 3D input features is that a certain degradation phenomenon will occur if the conditions are specific enough (the conditions make specific degradation phenomena thermodynamically feasible). Providing the model with information about the time spent in these conditions allows it to implicitly deduce the kinetics of the reactions contributing to capacity loss. Including additional features, such as the integral of current, expands the range of conditions considered by the models, as variations within a condition can now be accounted for. Figure 9 depicts the thresholds established for each of the stress factors across both databases. In all three neural networks, the output of the machine learning model is multiplied by the hyperbolic tangent of the time in days, divided by two (this mathematical expression is illustrated in Figures 6–8). This step guarantees that the capacity loss for a fresh cell is exactly zero, ensuring that after a few days, the model's output aligns with that of the ML model while maintaining a smooth transition between these two states. The thresholds identified for the databases are presented in Figure 9.

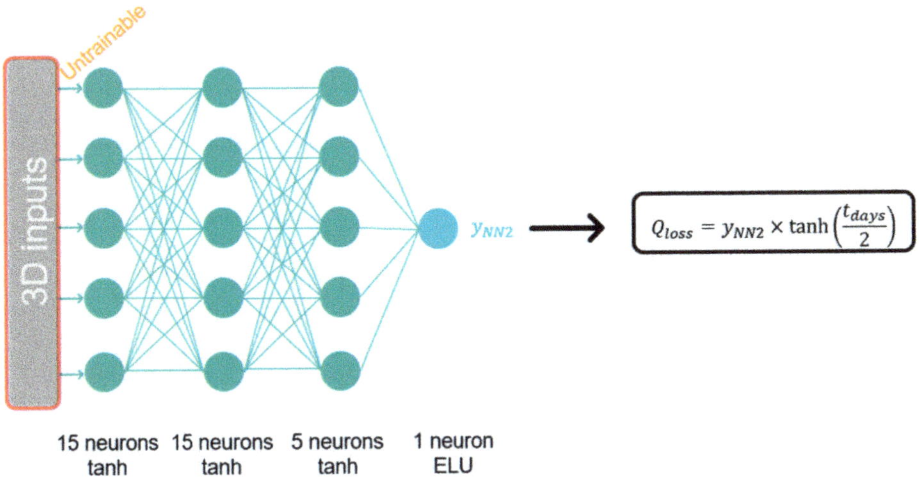

Figure 7. Neural network architecture 2 (NN_2): the model with an untrained layer.

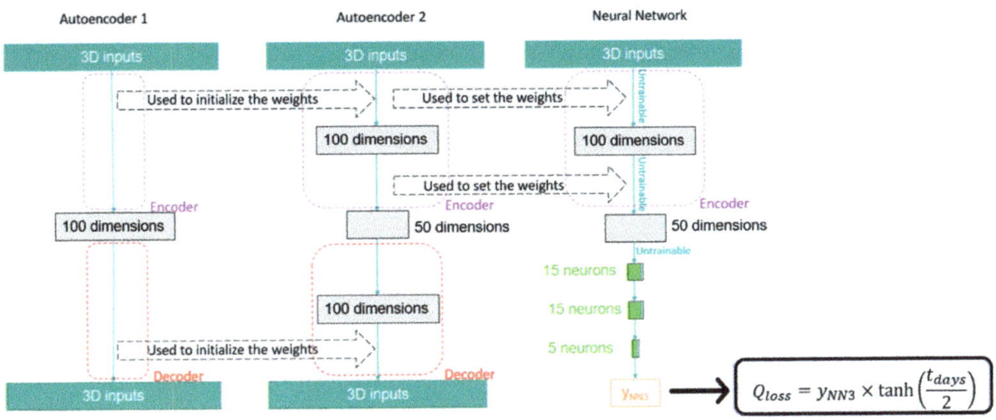

Figure 8. Neural network architecture 3 (NN_3): the model with the encoder. The activation functions of the green layers were the hyperbolic tangent.

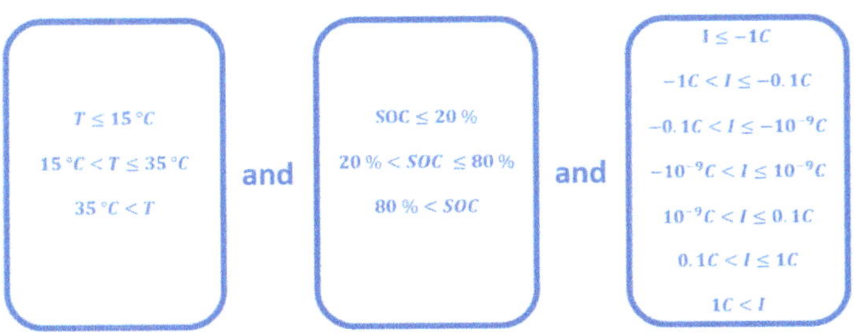

Figure 9. The thresholds applied in this study.

As mentioned above, the three variants of the method first concentrated on using thresholds to discretize the stress factors, which are real values that change over time. To achieve this, specific thresholds were established, and models were constructed based on the time spent by the cell between these thresholds. The aim was to develop a general model; therefore, the thresholds were not selected based on the distribution of the measurements. You et al. [35] employed a clustering method to delineate their zones, while other researchers, such as Greenbank et al. [45], used the distribution of the stress factors to determine the thresholds. Von Bülow et al. selected the thresholds without considering the distribution of the measurements [39]. In the present study, the thresholds were also chosen arbitrarily to represent the conditions experienced by the cell. This approach resulted in the following inputs for the one-dimensional models, where negative currents indicate discharge currents (see also Figure 5):

1. Age of the cell
2. Time spent by the cell with low SOC: $\int_{SOC\leq 20} dt$
3. Time spent by the cell with medium SOC: $\int_{20<SOC\leq 80} dt$
4. Time spent by the cell with high SOC: $\int_{80<SOC} dt$
5. Time spent by the cell at a low temperature: $\int_{T\leq 15} dt$
6. Time spent by the cell at a medium temperature: $\int_{15<T\leq 35} dt$
7. Time spent by the cell at a high temperature: $\int_{35<T} dt$
8. Time spent by the cell with a high discharge current: $\int_{I\leq -1C} dt$
9. Time spent by the cell with a medium discharge current: $\int_{-1C<I\leq -0.1C} dt$
10. Time spent by the cell with a low discharge current: $\int_{-0.1C<I\leq -10^{-9}C} dt$
11. Time spent by the cell almost in the calendar: $\int_{-10^{-9}C<I\leq 10^{-9}C} dt$
12. Time spent by the cell with a low charge current: $\int_{10^{-9}C<I\leq 0.1C} dt$
13. Time spent by the cell with a medium charge current: $\int_{0.1C<I\leq 1C} dt$
14. Time spent by the cell with a high charge current: $\int_{1C<I} dt$

As an example, the second feature: $\int_{SOC\leq 20} dt$ corresponds to the time the cell Ha spent with a SOC below 20%. Features 3 to 14 should be understood in the same way. As only one parameter is used to build a zone with this method, this data transformation is referred to as 1D zones. In contrast, the 3D approaches rely on the three stress factors at the same time, as can be seen in the example below of five features that are used in the 3D models (see also Figures 5 and 9):

- $\int_{\tau=0}^{t} f(\tau)d\tau$ where $\begin{cases} f(\tau) = 1 \ if \ T \leq 15°C \ and \ SOC > 80\% \ and - 10^{-9}C < I \leq 10^{-9}C \\ f(\tau) = 0 \ else \end{cases}$
- $\int_{\tau=0}^{t} f(\tau)d\tau$ where $\begin{cases} f(\tau) = 1 \ if \ T > 35°C \ and \ SOC > 80\% \ and \ - 10^{-9}C < I \leq 10^{-9}C \\ f(\tau) = 0 \ else \end{cases}$
- $\int_{\tau=0}^{t} f(\tau)d\tau$ where $\begin{cases} f(\tau) = SOC \ if \ T \leq 15°C \ and \ SOC > 80\% \ and \ - 10^{-9}C < I \leq 10^{-9}C \\ f(\tau) = 0 \ else \end{cases}$
- $\int_{\tau=0}^{t} f(\tau)d\tau$ where $\begin{cases} f(\tau) = I \ if \ T \leq 15°C \ and \ SOC \leq 20\% \ and \ - 10^{-9}C < I \leq 10^{-9}C \\ f(\tau) = 0 \ else \end{cases}$
- $\int_{\tau=0}^{t} f(\tau)d\tau$ where $\begin{cases} f(\tau) = T \ if \ 15°C < T \leq 35°C \ and \ 20\% < SOC \leq 80\% \ and \ 1C < I \\ f(\tau) = 0 \ else \end{cases}$

The ML models illustrated in Figures 6–8 are characterized by their hyperparameters, which include the number of layers, the number of neurons per layer, or the activation functions employed. Hyperparameters differ from the weights and biases of the model because they remain unchanged during the model's training phase. Numerous techniques exist to select the set of hyperparameters that yields the best performance on a given dataset [36]. In the models presented in Section 4, the hyperparameters were selected using a trial-and-error method, which involved iteratively tuning the model's hyperparameters. The other techniques mentioned, like Bayesian optimization or the hyperband, are more automatic. Their core idea is to conduct meta-optimization to adapt the hyperparameters of a model to a given dataset. This is referred to as meta-optimization because it is the optimization of a process that includes an optimization (in the context of machine learning, the included optimization mentioned consists of training the models). In this work, the aim was to propose architectures that are general enough to be used on any database, so they must be well-adapted to working with several databases. This task would, thus, consist of a multi-objective meta-optimization problem. Multi-objective optimization often results in a set of optimal solutions, known as a Pareto set. Having this goal in mind, the trial-and-error approach was chosen because it enables the authors to see the different optima and choose the best trade-off, rather than having an automatic method that yields only one optimum. All models presented in this work were trained and utilized on a Dell Precision 7530 laptop, which was equipped with an Intel Core i7 8850H CPU. The training durations for these models were under 10 s, while the calculations of the capacity losses on the training data took less than 0.2 s for all models. Due to the data augmentation technique implemented for the autoencoders, their training durations were longer, this being approximately 45 s for the EVERLASTING dataset and around 25 s for the Bills dataset.

These model architectures and these model inputs could be used to estimate other battery-aging-related information such as the internal resistance, the remaining useful life, or even the loss of active materials and the loss of lithium inventory. In the present work, only the capacity loss was estimated.

4. Results and Discussion

For this section, the models were developed, and their results were then compared using performance metrics for evaluation. The metrics utilized were the mean absolute error (*MAE*), the root mean squared error (*RMSE*), and the correlation coefficient (R^2). Both the *MAE* and *RMSE* were expressed in the same units as capacity loss, shown specifically as a percentage of initial capacity. The mathematical definitions of these metrics are as follows:

$$MAE = \sum_{k=1}^{N} \frac{|y_k - \hat{y_k}|}{N} \tag{11}$$

$$RMSE = \sqrt{\sum_{k=1}^{N} \frac{(y_k - \hat{y_k})^2}{N}} \tag{12}$$

$$R^2 = 1 - \frac{\sum_{k=1}^{N} (y_k - \hat{y_k})^2}{\sum_{k=1}^{N} (y_k - \overline{y})^2} \tag{13}$$

where N is the number of Q_{loss} measurements, y is the measured Q_{loss}, $\hat{y}$ is the estimated Q_{loss}, and $\overline{y}$ is the average of all the Q_{loss} measurements on which R^2 is calculated. These metrics are presented for both the training and the testing datasets. The objective was to develop models that are capable of predicting battery aging under conditions that differ from those in the training data. Therefore, only the metrics from the test data were utilized for model comparison.

4.1. Aging Models on the EVERLASTING Dataset

For this section, the performance of the models on the EVERLASTING database was assessed. The results of the different models are presented in Figures 10 and 11. Figure 10 shows the party plots of the different models. The x-axes of these plots represent the capacity loss measurements, and the y-axes represent the corresponding capacity loss estimated by the model. The identity function $y = x$ is represented on the graphs because, with a perfect model, all points would be on that line. Figure 11 depicts the metrics associated with each model using the EVERLASTING data. Using this database, the 3D model with an untrained layer (NN_2) outperformed all other models, achieving a value for $R^2{}_{test} = 0.9$. The other two ML models yielded similar results, which were superior to those of the empirical models. The commercial model was calibrated using the calendar and constant current data from the training database, as it could not be optimized for driving profiles. Its results surpassed those of the basic model. As expected, the basic model exhibited poor predictive capability; the parity plot revealed horizontal lines where the model prediction yielded similar values, corresponding to the capacity measurements taken for cells of similar ages, despite their differing capacity losses. In Figure 10d, the commercial model shown displayed more test data points than the other figures because the other models were trained on driving profiles. Given the differing conditions between the training and the testing sets, the models' ability to generalize was evaluated. Consequently, the ML models performed slightly worse on the test data than on the training data, although their overall performance demonstrates that they effectively captured the degradation of the cells. The empirical models were more constrained, possessing a predefined aging profile and fewer degrees of freedom. The authors assume that this explains the slight improvement observed in the commercial model between the training and testing data.

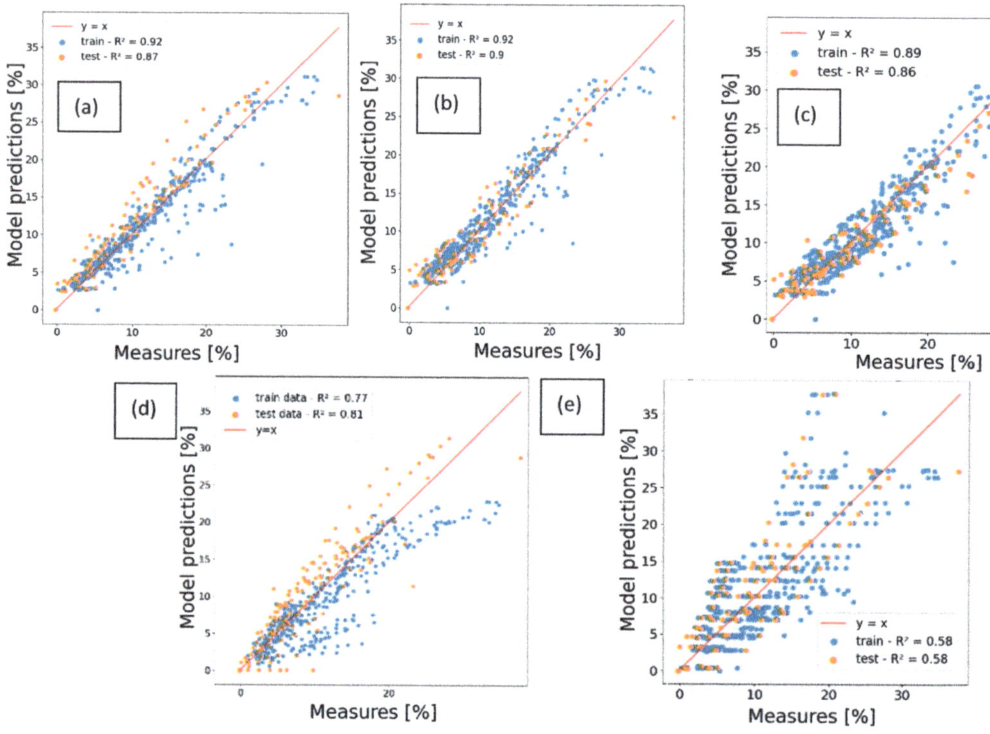

Figure 10. Parity plots of the capacity loss of (**a**) the 1D model (NN_1), (**b**) the model with an untrained layer (NN_2), (**c**) the model with the auto-encoders (NN_3), (**d**) the commercial model, and (**e**) the basic model on the EVERLASTING data.

4.2. Aging Models on the Bills Dataset

The models described in Sections 3.2 and 3.3 were implemented on the Bills et al. database. These models were applied to the dataset using the cells VAH02, VAH10, VAH15, VAH22, and VAH28 as test data, while the remaining 17 cells served as training data. The resulting parity plots are presented in Figure 12 and the corresponding performance metrics are shown in Figure 13. The results indicate that all models exhibited an R^2_{test} value between 0.92 and 0.96, suggesting similar performance and good approximations of capacity loss within this database. However, the model utilizing the auto-encoder performed slightly worse on both the training and test sets compared to the other two ML models. Although the differences were minor, the NN_1 model demonstrated the best performance on both training and test data. Notably, NN_1 was the only ML model for which performance did not decrease when generalizing to the test data. In contrast, the empirical models demonstrated a slight improvement in performance from the training data to the test data.

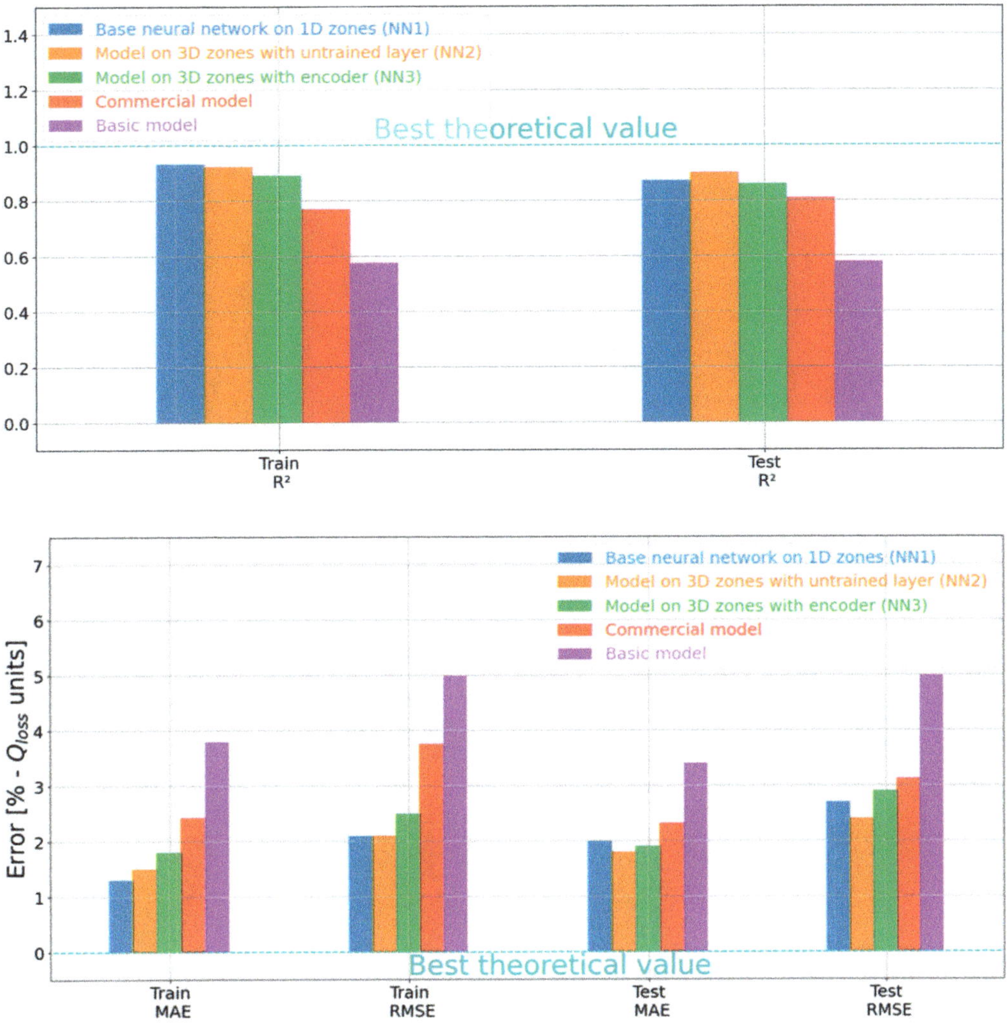

Figure 11. Comparison of the performance metrics for the different models on the EVERLASTING database.

All models demonstrated lower performance on the EVERLASTING data compared to the Bills data. This outcome was to be expected as the EVERLASTING database exhibits greater variability in the graph of Q_{loss} versus time than the Bills database (see Figures 2 and 4). This variability also explains the similar performance of all models on the Bills data. Consequently, all models yielded comparable results on a less diverse database, while the proposed machine learning models achieved better performance on the more diverse EVERLASTING dataset. This suggests that the machine learning models possess a superior ability to capture battery degradation performance.

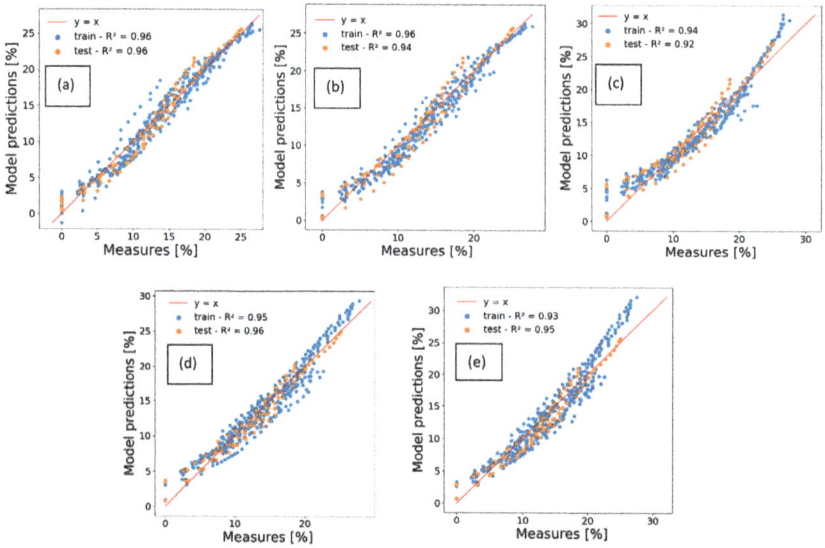

Figure 12. Parity plots of the capacity loss of (**a**) the 1D model (NN_1), (**b**) the model with an untrained layer (NN_2), (**c**) the model with the auto-encoders (NN_3), (**d**) the commercial model, and (**e**) the basic model on the Bills data.

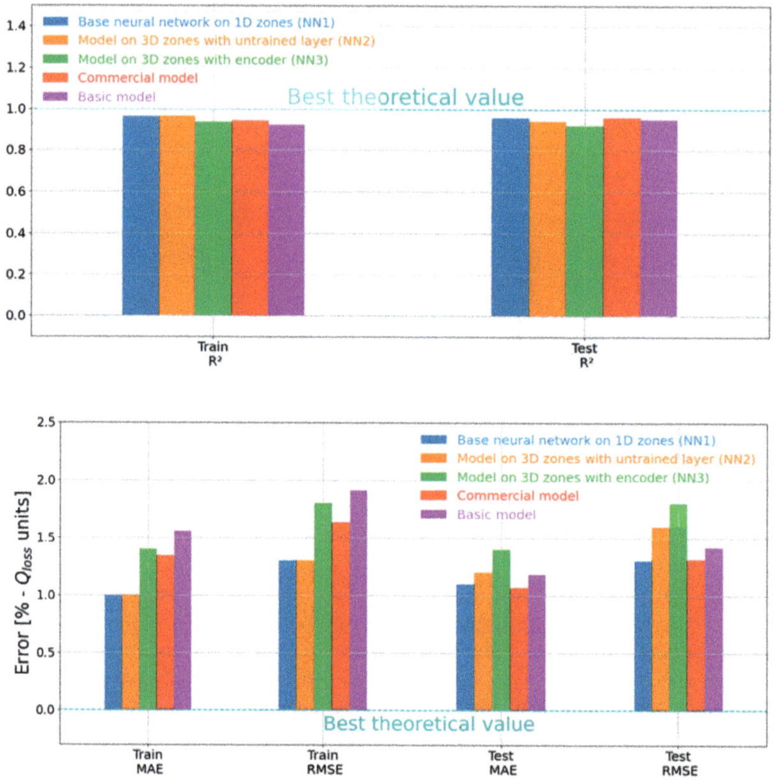

Figure 13. Histogram of the performances of the models on the Bills dataset.

4.3. Overall Comparison of the Models

A metric can be derived from the results obtained from the test data in both databases. This metric is the total RMSE. The definition of this metric $RMSE_{tot}$, and its relationship with the previously mentioned RMSE is shown in the equations:

$$RMSE_{tot} = \sqrt{\frac{\sum_{all\ test\ measurements}(y - \hat{y})^2}{N_{tot}}} \tag{14}$$

$$RMSE_{tot} = \sqrt{\frac{\sum_{Bills\ test}(y - \hat{y})^2 + \sum_{EVERLASTING\ test}(y - \hat{y})^2}{N_{tot}}} \tag{15}$$

$$RMSE_{tot} = \sqrt{\frac{RMSE_1^2 \times N_1 + RMSE_2^2 \times N_2}{N_1 + N_2}} \tag{16}$$

In these equations, y is the measured Q_{loss}, $\hat{y}$ is the estimated Q_{loss}, N_{tot} is the total number of capacity losses considered, $RMSE_1$ is the RMSE of the test data from Bills, $RMSE_2$ is the $RMSE$ of the test data from EVERLASTING, N_1 is the number of capacity losses in the test data from Bills, N_2 is the number of capacity losses in the test data from EVERLASTING.

When applied to a model, this metric corresponds to the precision of the model in every instance where it has been tested. This overall score of the model enables its comparison. Figure 14 shows a comparison of the five models considered in this study. In the figure, it can be observed that the best-ranking model is the model with the untrained layer (NN_2), inspired by extreme learning, closely followed by the 1D model (NN_1). The 3D model involving the autoencoder (NN_3) is less accurate, although it is more accurate than the commercial model. The basic model is ranked last.

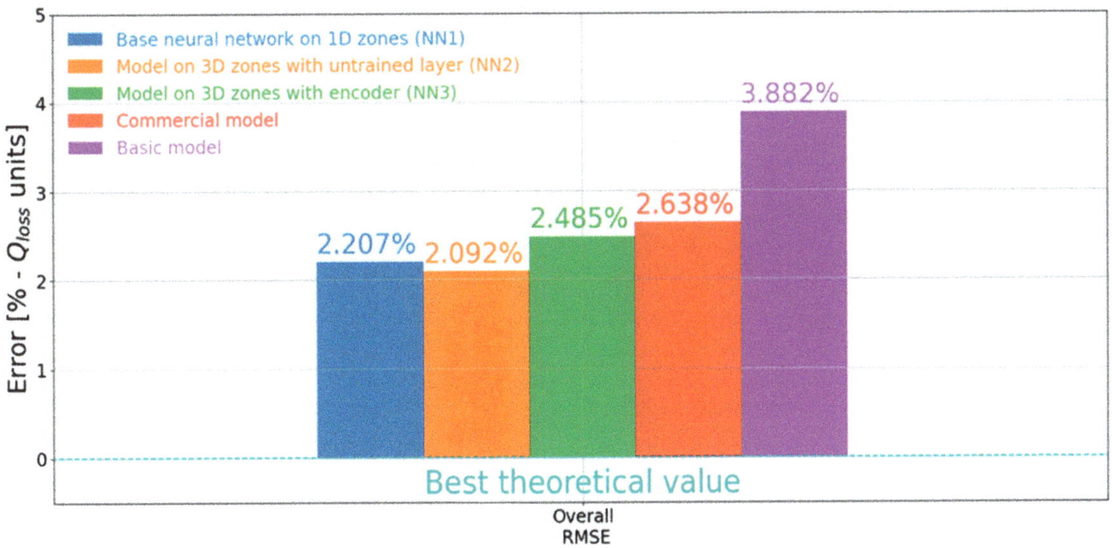

Figure 14. Overall model comparison.

A possible explanation for the results obtained from each of the proposed machine learning models is that the 3D inputs contain more information relevant to understanding battery aging than the 1D inputs. This discrepancy may account for the superior performance of NN_2 compared to NN_1, particularly since 1D inputs can be derived from 3D inputs, while the reverse is not feasible. To explain the lower performance of NN_3, one hypothesis is that the significant variance in the aging data shared by the encoder with the network is because it does not include all the information from the 3D inputs related to capacity loss, nor does it include all the information used in NN_2 to calculate the capacity loss. Another potential explanation is that this variance is compressed into too few variables, preventing the network following the encoder from accurately interpreting the information drawn from the latent space.

5. Conclusions

In conclusion, three machine learning frameworks were presented and applied to two distinct datasets. When used without modifications, these frameworks delivered good results and demonstrated robustness, highlighting their reliability. The proposed models outperformed the commercial model on the EVERLASTING dataset. In contrast, all models exhibited similar performances on the Bills dataset. When calculating the overall RMSE across all test data points from both databases, the one-dimensional model and the model inspired by an extreme learning machine performed best, followed by the other 3D model, which outperformed the two existing methods: the commercial model and the basic model. The basic model had the largest overall error.

The proposed method for constructing zones in the battery's stress factor space considers the integral over time of several functions as the inputs for machine learning models to predict battery degradation. This method has been validated on two databases, using the same thresholds to separate the stress factors. Such general approaches have not been applied to these datasets in the existing literature. In addition, novel features were incorporated, autoencoders were utilized, and the obtained results were compared with those of existing models, which activity distinguishes this work from previous studies. The datasets included various temperatures and aging profiles, such as calendar aging, constant current cycling data, and profiles inspired by real-world scenarios, including driving profiles and eVTOL missions. The proposed ML approaches yielded competitive or superior results compared to the commercial model on both datasets, demonstrating their versatility.

Several perspectives can be considered for this work. These include predicting the increase in internal resistance of the cells, assessing the loss of active material at each electrode, as well as the loss of lithium inventory, alongside capacity loss. All these assessments utilize the same model inputs and architectures. Additionally, another perspective is to compare multilayer perceptrons with recurrent neural networks.

Author Contributions: Conceptualization, Q.M., G.D., A.L., R.M. and P.V.; data curation, Q.M.; methodology, Q.M.; software, Q.M.; validation, Q.M.; formal analysis, Q.M.; investigation, Q.M.; writing—original draft preparation, Q.M.; writing—review and editing, G.D., A.L., R.M. and P.V.; visualization, Q.M.; supervision, G.D., A.L., R.M. and P.V.; project administration, Q.M., G.D., A.L., R.M. and P.V.; funding acquisition, G.D., A.L., R.M. and P.V. All authors have read and agreed to the published version of the manuscript.

Funding: This research was funded by Siemens Industries Software, with subsidies from the Association Nationale de la Recherche et de la Technologie (ANRT) under a CIFRE convention.

Data Availability Statement: Data from the Bills dataset are available here: https://figshare.com/articles/dataset/eVTOL_Battery_Dataset/14226830 (accessed on 13 October 2024). Data from the EVERLASTING dataset are accessible following the links: https://data.4tu.nl/datasets/e42bca59-f1dd-495a-92c9-8b01d6b64040 (accessed on 13 October 2024) https://data.4tu.nl/datasets/e19fe272-4f46-450c-9125-6545c4c1a98b (accessed on 13 October 2024), https://data.4tu.nl/collections/EVERLASTING_Electric_Vehicle_Enhanced_Range_Lifetime_And_Safety_Through_INGenious_battery_management_/5065445 (accessed on 13 October 2024). Data can also be made available upon request.

Conflicts of Interest: Authors Quentin Mayemba and An Li were employed by the company Siemens Digital Industries Software. The remaining authors declare that the research was conducted in the absence of any commercial or financial relationships that could be construed as a potential conflict of interest.

References

1. Peng, J.; Meng, J.; Chen, D.; Liu, H.; Hao, S.; Sui, X.; Du, X. A Review of Lithium-Ion Battery Capacity Estimation Methods for Onboard Battery Management Systems: Recent Progress and Perspectives. *Batteries* **2022**, *8*, 229. [CrossRef]
2. Mayemba, Q.; Mingant, R.; Li, A.; Ducret, G.; Venet, P. Aging datasets of commercial lithium-ion batteries: A review. *J. Energy Storage* **2024**, *83*, 110560. [CrossRef]
3. University of Liverpool Lithium Cobalt Oxide–LiCoO2–Conduction Animation. Available online: https://www.chemtube3d.com/lib_lco-2/ (accessed on 10 April 2024).
4. Maher, K.; Yazami, R. A study of lithium ion batteries cycle aging by thermodynamics techniques. *J. Power Sources* **2014**, *247*, 527–533. [CrossRef]
5. McBrayer, J.D.; Rodrigues, M.-T.F.; Schulze, M.C.; Abraham, D.P.; Apblett, C.A.; Bloom, I.; Carroll, G.M.; Colclasure, A.M.; Fang, C.; Harrison, K.L.; et al. Calendar aging of silicon-containing batteries. *Nat. Energy* **2021**, *6*, 866–872. [CrossRef]
6. Pelletier, S.; Jabali, O.; Laporte, G.; Veneroni, M. Battery degradation and behaviour for electric vehicles: Review and numerical analyses of several models. *Transp. Res. Part B Methodol.* **2017**, *103*, 158–187. [CrossRef]
7. Fath, J.P.; Dragicevic, D.; Bittel, L.; Nuhic, A.; Sieg, J.; Hahn, S.; Alsheimer, L.; Spier, B.; Wetzel, T. Quantification of aging mechanisms and inhomogeneity in cycled lithium-ion cells by differential voltage analysis. *J. Energy Storage* **2019**, *25*, 100813. [CrossRef]
8. Ahn, Y.; Jo, Y.N.; Cho, W.; Yu, J.-S.; Kim, K.J. Mechanism of Capacity Fading in the $LiNi_{0.8}Co_{0.1}Mn_{0.1}O_2$ Cathode Material for Lithium-Ion Batteries. *Energies* **2019**, *12*, 1638. [CrossRef]
9. Chen, C.-F.; Barai, P.; Mukherjee, P.P. An overview of degradation phenomena modeling in lithium-ion battery electrodes. *Curr. Opin. Chem. Eng.* **2016**, *13*, 82–90. [CrossRef]
10. Reniers, J.M.; Mulder, G.; Howey, D.A. Review and Performance Comparison of Mechanical-Chemical Degradation Models for Lithium-Ion Batteries. *J. Electrochem. Soc.* **2019**, *166*, A3189–A3200. [CrossRef]
11. Rowden, B.; Garcia-Araez, N. A review of gas evolution in lithium ion batteries. *Energy Rep.* **2022**, *6*, 10–18. [CrossRef]
12. Kabir, M.M.; Demirocak, D.E. Degradation mechanisms in Li-ion batteries: A state-of-the-art review. *Int. J. Energy Res.* **2017**, *41*, 1963–1986. [CrossRef]
13. Gao, H.; Yan, Q.; Holoubek, J.; Yin, Y.; Bao, W.; Liu, H.; Baskin, A.; Li, M.; Cai, G.; Li, W.; et al. Enhanced Electrolyte Transport and Kinetics Mitigate Graphite Exfoliation and Li Plating in Fast-Charging Li-Ion Batteries. *Adv. Energy Mater.* **2023**, *13*, 2202906. [CrossRef]
14. Winter, M.; Barnett, B.; Xu, K. Before Li Ion Batteries. *Chem. Rev.* **2018**, *118*, 11433–11456. [CrossRef] [PubMed]
15. Zhang, Z.; Yang, J.; Huang, W.; Wang, H.; Zhou, W.; Li, Y.; Li, Y.; Xu, J.; Huang, W.; Chiu, W.; et al. Cathode-Electrolyte Interphase in Lithium Batteries Revealed by Cryogenic Electron Microscopy. *Matter* **2021**, *4*, 302–312. [CrossRef]
16. Guo, L.; Thornton, D.B.; Koronfel, M.A.; Stephens, I.E.L.; Ryan, M.P. Degradation in lithium ion battery current collectors. *J. Phys. Energy* **2021**, *3*, 32015. [CrossRef]
17. Vetter, J.; Novák, P.; Wagner, M.R.; Veit, C.; Möller, K.-C.; Besenhard, J.O.; Winter, M.; Wohlfahrt-Mehrens, M.; Vogler, C.; Hammouche, A. Ageing mechanisms in lithium-ion batteries. *J. Power Sources* **2005**, *147*, 269–281. [CrossRef]
18. Chae, S.; Kim, N.; Ma, J.; Cho, J.; Ko, M. One-to-One Comparison of Graphite-Blended Negative Electrodes Using Silicon Nanolayer-Embedded Graphite versus Commercial Benchmarking Materials for High-Energy Lithium-Ion Batteries. *Adv. Energy Mater.* **2017**, *7*, 15. [CrossRef]
19. Lin, C.; Tang, A.; Mu, H.; Wang, W.; Wang, C. Aging Mechanisms of Electrode Materials in Lithium-Ion Batteries for Electric Vehicles. *J. Chem.* **2015**, *2015*, 104673. [CrossRef]
20. Yang, R.; Yu, G.; Wu, Z.; Lu, T.; Hu, T.; Liu, F.; Zhao, H. Aging of lithium-ion battery separators during battery cycling. *J. Energy Storage* **2023**, *63*, 107107. [CrossRef]
21. Wheeler, W.; Venet, P.; Bultel, Y.; Sari, A.; Riviere, E. Aging in First and Second Life of G/LFP 18650 Cells: Diagnosis and Evolution of the State of Health of the Cell and the Negative Electrode under Cycling. *Batteries* **2024**, *10*, 137. [CrossRef]
22. Abada, S.; Petit, M.; Lecocq, A.; Marlair, G.; Sauvant-Moynot, V.; Huet, F. Combined experimental and modeling approaches of the thermal runaway of fresh and aged lithium-ion batteries. *J. Power Sources* **2018**, *399*, 264–273. [CrossRef]
23. O'Kane, S.E.J.; Ai, W.; Madabattula, G.; Alonso-Alvarez, D.; Timms, R.; Sulzer, V.; Edge, J.S.; Wu, B.; Offer, G.J.; Marinescu, M. Lithium-ion battery degradation: How to model it. *Phys. Chem. Chem. Phys.* **2022**, *24*, 7909–7922. [CrossRef]
24. Lopetegi, I.; Plett, G.L.; Trimboli, M.S.; Yeregui, J.; Oca, L.; Rojas, C.; Miguel, E.; Iraola, U. Lithium-ion Battery Aging Prediction with Electrochemical Models: P2D vs. SPMe. In Proceedings of the 2023 IEEE Vehicle Power and Propulsion Conference (VPPC), Milan, Italy, 24–27 October 2023; IEEE: Piscataway, NJ, USA, 2023; pp. 1–7.
25. Mingant, R.; Petit, M.; Belaïd, S.; Bernard, J. Data-driven model development to predict the aging of a Li-ion battery pack in electric vehicles representative conditions. *J. Energy Storage* **2021**, *39*, 102592. [CrossRef]
26. Vermeer, W.; Chandra Mouli, G.R.; Bauer, P. A Comprehensive Review on the Characteristics and Modeling of Lithium-Ion Battery Aging. *IEEE Trans. Transp. Electrif.* **2022**, *8*, 2205–2232. [CrossRef]
27. Bülow, F.; von Meisen, T. A review on methods for state of health forecasting of lithium-ion batteries applicable in real-world operational conditions. *J. Energy Storage* **2023**, *57*, 105978. [CrossRef]
28. Sui, X.; He, S.; Vilsen, S.B.; Meng, J.; Teodorescu, R.; Stroe, D.-I. A review of non-probabilistic machine learning-based state of health estimation techniques for Lithium-ion battery. *Appl. Energy* **2021**, *300*, 117346. [CrossRef]

29. Guo, W.; Sun, Z.; Vilsen, S.B.; Meng, J.; Stroe, D.I. Review of "grey box" lifetime modeling for lithium-ion battery: Combining physics and data-driven methods. *J. Energy Storage* **2022**, *56*, 105992. [CrossRef]
30. Khaleghi, S.; Hosen, M.S.; van Mierlo, J.; Berecibar, M. Towards machine-learning driven prognostics and health management of Li-ion batteries. A comprehensive review. *Renew. Sustain. Energy Rev.* **2024**, *192*, 114224. [CrossRef]
31. Cao, L.; Xu, R.; Bi, Y. Research on Life Prediction of Lithium-ion Battery based on WEMD-ARIMA Model. In Proceedings of the 2022 34th Chinese Control and Decision Conference (CCDC), Hefei, China, 15–17 August 2022; IEEE: Piscataway, NJ, USA, 2022; pp. 2774–2779.
32. Jorge, I.; Mesbahi, T.; Samet, A.; Boné, R. Time Series Feature extraction for Lithium-Ion batteries State-Of-Health prediction. *J. Energy Storage* **2023**, *59*, 106436. [CrossRef]
33. Lombardo, T.; Duquesnoy, M.; El-Bouysidy, H.; Årén, F.; Gallo-Bueno, A.; Jørgensen, P.B.; Bhowmik, A.; Demortière, A.; Ayerbe, E.; Alcaide, F.; et al. Artificial Intelligence Applied to Battery Research: Hype or Reality? *Chem. Rev.* **2022**, *122*, 10899–10969. [CrossRef]
34. Nuhic, A.; Terzimehic, T.; Soczka-Guth, T.; Buchholz, M.; Dietmayer, K. Health diagnosis and remaining useful life prognostics of lithium-ion batteries using data-driven methods. *J. Power Sources* **2013**, *239*, 680–688. [CrossRef]
35. You, G.; Park, S.; Oh, D. Real-time state-of-health estimation for electric vehicle batteries: A data-driven approach. *Appl. Energy* **2016**, *176*, 92–103. [CrossRef]
36. Richardson, R.R.; Osborne, M.A.; Howey, D.A. Battery health prediction under generalized conditions using a Gaussian process transition model. *J. Energy Storage* **2019**, *23*, 320–328. [CrossRef]
37. Song, L.; Zhang, K.; Liang, T.; Han, X.; Zhang, Y. Intelligent state of health estimation for lithium-ion battery pack based on big data analysis. *J. Energy Storage* **2020**, *32*, 101836. [CrossRef]
38. Greenacre, M.; Groenen, P.J.F.; Hastie, T.; D'Enza, A.I.; Markos, A.; Tuzhilina, E. Principal component analysis. *Nat. Rev. Methods Primers* **2020**, *2*, 100. [CrossRef]
39. Bülow, F.; von Mentz, J.; Meisen, T. State of health forecasting of Lithium-ion batteries applicable in real-world operational conditions. *J. Energy Storage* **2021**, *44*, 103439. [CrossRef]
40. Severson, K.A.; Attia, P.M.; Jin, N.; Perkins, N.; Jiang, B.; Yang, Z.; Chen, M.H.; Aykol, M.; Herring, P.K.; Fraggedakis, D.; et al. Data-driven prediction of battery cycle life before capacity degradation. *Nat. Energy* **2019**, *4*, 383–391. [CrossRef]
41. Yang, L.; Shami, A. On hyperparameter optimization of machine learning algorithms: Theory and practice. *Neurocomputing* **2020**, *415*, 295–316. [CrossRef]
42. Zhang, Y.; Wik, T.; Bergström, J.; Pecht, M.; Zou, C. A machine learning-based framework for online prediction of battery ageing trajectory and lifetime using histogram data. *J. Power Sources* **2022**, *526*, 231110. [CrossRef]
43. Attia, P.M.; Grover, A.; Jin, N.; Severson, K.A.; Markov, T.M.; Liao, Y.-H.; Chen, M.H.; Cheong, B.; Perkins, N.; Yang, Z.; et al. Closed-loop optimization of fast-charging protocols for batteries with machine learning. *Nature* **2020**, *578*, 397–402. [CrossRef]
44. NASA. Available online: https://papers.phmsociety.org/index.php/phmconf/article/view/2490 (accessed on 3 April 2024).
45. Greenbank, S.; Howey, D.A. Piecewise-linear modelling with automated feature selection for Li-ion battery end-of-life prognosis. *Mech. Syst. Signal Process.* **2023**, *184*, 109612. [CrossRef]
46. Attia, P.M.; Bills, A.; Brosa Planella, F.; Dechent, P.; dos Reis, G.; Dubarry, M.; Gasper, P.; Gilchrist, R.; Greenbank, S.; Howey, D.; et al. Review—"Knees" in Lithium-Ion Battery Aging Trajectories. *J. Electrochem. Soc.* **2022**, *169*, 60517. [CrossRef]
47. Trad, K.; Govindarajan, J. *D2.3-Report Containing Aging Test Profiles and Test Results*; EVERLASTING: Yakima, WA, USA, 2020; Available online: https://everlasting-project.eu/wp-content/uploads/2020/03/EVERLASTING_D2.3_final_20200228.pdf (accessed on 13 October 2024).
48. Bills, A.; Viswanathan, V.; Sripad, S.; Frank, E.; Charles, D.; Fredericks, W.L. *eVTOL Battery Dataset*; Carnegie Mellon University: Pittsburgh, PA, USA, 2023.
49. dos Reis, G.; Strange, C.; Yadav, M.; Li, S. Lithium-ion battery data and where to find it. *Energy AI* **2021**, *5*, 100081. [CrossRef]
50. Hassini, M.; Redondo-Iglesias, E.; Venet, P. Lithium–Ion Battery Data: From Production to Prediction. *Batteries* **2023**, *9*, 385. [CrossRef]
51. Yang, X.-G.; Liu, T.; Ge, S.; Rountree, E.; Wang, C.-Y. Challenges and key requirements of batteries for electric vertical takeoff and landing aircraft. *Joule* **2021**, *5*, 1644–1659. [CrossRef]
52. CC BY. Available online: https://creativecommons.org/licenses/by/4.0/ (accessed on 13 October 2024).
53. Siemens Digital Industries Software. *Simcenter AMESim V2310 (Advanced Modeling Environment for Performing Simulations)*; Siemens Digital Industries Software: Plano, TX, USA, 2023.
54. Jafari, M.; Khan, K.; Gauchia, L. Deterministic models of Li-ion battery aging: It is a matter of scale. *J. Energy Storage* **2018**, *20*, 67–77. [CrossRef]
55. Huang, G.-B.; Zhu, Q.-Y.; Siew, C.-K. Extreme learning machine: Theory and applications. *Neurocomputing* **2006**, *70*, 489–501. [CrossRef]
56. Wang, J.; Lu, S.; Wang, S.-H.; Zhang, Y.-D. A review on extreme learning machine. *Multimed. Tools Appl.* **2022**, *81*, 41611–41660. [CrossRef]

57. Michelucci, U. An Introduction to Autoencoders. *arXiv* **2022**, arXiv:2201.03898.
58. Li, P.; Pei, Y.; Li, J. A comprehensive survey on design and application of autoencoder in deep learning. *Appl. Soft Comput.* **2023**, *138*, 110176. [CrossRef]

batteries

MDPI

Article

Learning the Ageing Behaviour of Lithium-Ion Batteries Using Electric Vehicle Fleet Analysis

Thomas Lehmann *, Erik Berendes, Richard Kratzing and Gautam Sethia

Fraunhofer Institute for Transportation and Infrastructure Systems, 01069 Dresden, Germany; gautam.sethia@ivi.fraunhofer.de (G.S.)
* Correspondence: thomas.lehmann@ivi.fraunhofer.de

Abstract: This article presents the results of the Febal project, where the aim was to parametrize a stress-factor-based ageing model for Lithium-ion batteries using operation data of an electric fleet. Contrary to state-of-the-art methods, this approach does not rely on laboratory ageing tests only. Instead, a novel physics-informed learning procedure is used to combine the accuracy and flexibility of data-driven approaches with the extrapolation properties of physical models. The ageing model is parameterized in a two-stage process. In order to cover data ranges not present in operation, a laboratory ageing test campaign is used as a baseline. In the second stage, transfer learning is used to adjust a subset of the model parameters to fit data of different cells. This approach is not only applied to laboratory measurements but also validated by a series of capacity checkup tests performed with a fleet of electric vehicles. Results show the improved state-of-health (SOH) prediction of the proposed model parameterization method.

Keywords: field data; informed neural network; battery ageing

1. Introduction

The performance degradation of lithium-ion cells, especially within the context of automotive applications, poses significant challenges that are closely studied by both researchers and industry stakeholders. Accurate prediction of the ageing behaviour of lithium-ion batteries is essential for improving the performance, safety, and longevity of electric vehicles (EVs). As such, numerous research efforts have been dedicated to understanding and modelling the factors that contribute to battery degradation over time. The prediction of the ageing behaviour of lithium-ion batteries is addressed in various research activities, an overview of which is given in [1]. Most of these studies deal with laboratory data and often focus on specific ageing effects. Semi-empirical functions are commonly used to fit ageing test data. A comprehensive overview of empirical ageing models can be found in [2,3]. The generalization of these models and the application to other cell types is studied in [4]. As a result, it can be stated that the optimal model structure depends on cell type as well as on the operation profile.

Recently, the detailed investigation of ageing modes has gained a lot of attention in research [5–8], where attempts to estimate cell intrinsic/physical states are carried out. This leads to deeper understanding of ageing states and enhances ageing prediction. Neural networks have also proven to be effective in several state-of-health (SOH) prediction scenarios. In [9], a combination of a convolution neural network (CNN) and a long short-term memory (LSTM) neural network is used to predict accelerated ageing. In [10], the application of bidirectional LSTM networks for capacity prediction is analysed.

Regardless of the model approach, comprehensive datasets of aged batteries under various operation conditions are needed for accurate parameterization. In [11], an overview of the design of an experiment as well as its application to experimental battery ageing investigation is given. Based on the presented theory, several ageing test campaigns are also evaluated. Unfortunately, open public datasets are very rare (in [12] an overview of available datasets is given). The transferability of laboratory experiments to field data is stated as one of the key challenges in battery state estimation in [13]. On the one hand, some causes of ageing that are relevant in practice, such as mechanical stress and imbalances in cells, have not been adequately investigated in the laboratory [14]. On the other hand, some conditions (e.g., extreme temperature effects) rarely occur in field data and should therefore be tested in the laboratory instead. In order to parameterize ageing models from multiple data sources or to apply models to other cell types, transfer learning methods can be used. An overview of the recent progress in battery state estimation using transfer learning is provided in [15]. In [16], transfer learning is used for calendric ageing prediction of different cell types. In [17], a cycle synchronization method is proposed in order to transfer ageing state information between cells. Most of the recent studies still focus on laboratory data only. In [18], a neural network model is trained with data derived from an operation. In [19], SOH estimation is performed using datasets from various electric vehicles. However, the combination of field and laboratory data is barely investigated in the literature.

In this article, a novel hybrid learning approach, developed within the Febal project [20], is proposed in order to make optimal use of both data sources, laboratory data and operational data. Herein, aspects from physics-informed neural networks as well as from transfer learning are combined (see Section 2.2.1). An empirical ageing model analogous to [21] is coupled to a feedforward neural network to enhance prediction accuracy (see Section 2.2). The aim is to guarantee good extrapolation properties while also preserving the flexibility and prediction accuracy of neural networks. A fleet of electric buses was monitored by us for several years and checked using regular capacity tests. The data were transferred in a specifically designed monitoring system. Furthermore, laboratory ageing tests of cells with similar cell chemistry were analysed in our lab facility (see Section 2.1). Our approach estimates stress maps, mapping operation conditions to ageing effects using laboratory data (see Section 2.3.1). Customized transfer learning approaches are thereafter applied to predict the ageing of other cell chemistries (see Section 3.2) and the electric vehicle fleet (see Section 3.3). The results show the superiority of the proposed hybrid model approach to commonly used empirical relationships and highlights the opportunities of using field data for model parameterization.

2. Materials and Methods

2.1. Datasets

The four datasets summarized in Table 1 are used in this study. There are three laboratory ageing campaigns and a dataset recorded from a fleet of electrical buses. The laboratory ageing data primarily consist of constant current experiments under fixed conditions. This means temperature, SOC range, and current rates are kept fixed. In the field data, these values are inherently dynamic. Another difference is the number of conducted checkup tests. Whereas in laboratory tests all cells are regularly examined in dedicated checkup tests, this cannot be realized in real electric vehicle fleets due to high investment of time and effort. Here, only some vehicles have undergone a dedicated charge and discharge procedure from which "ground truth" capacity values can be determined.

Exemplary checkup results of the calendric experiments of the Sony campaign are shown in Figure 1. The decline in capacity follows the same qualitative square-root-like pattern but with different ageing rates. In the other laboratory campaigns, linear ageing trends are also observed.

Figure 2 shows the range of important battery parameters for each bus of the entire bus fleet. It can be seen that the usage profiles are quite similar. Thermal management keeps the cells at comfortable temperatures during most of the operation phase, with some outliers at the start or during fast charging. The range of the SOC is also very similar. Buses are operated above 60% SOC most of the time. Charge rate and monthly full cycles differ based on the used charging station and driven route.

Table 1. Summary of the used datasets.

Name	Source	Operation Conditions of the Cell	Number of Cells/Vehicles
Sony	TUM [22]	• 5 °C–60 °C • SOC: 5–100%	97
batteries 2020	EU project [23]	• 25 °C–45 °C • SOC: 20–100%	146
Kokam	IVI (internal, unpublished)	• 10 °C–40 °C • SOC: 25–100%	28
vehicle fleet	transportation company	• 25 °C–30 °C • SOC: 40–100%	38

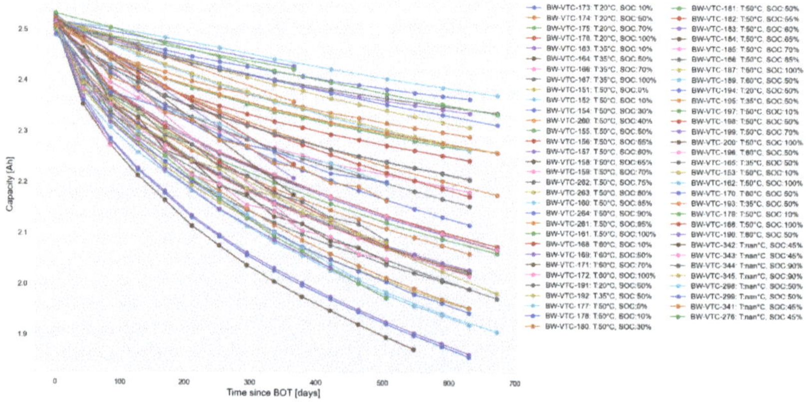

Figure 1. Determined capacities during calendric experiments of the Sony ageing test campaign as a function of time since beginning of testing (BOT).

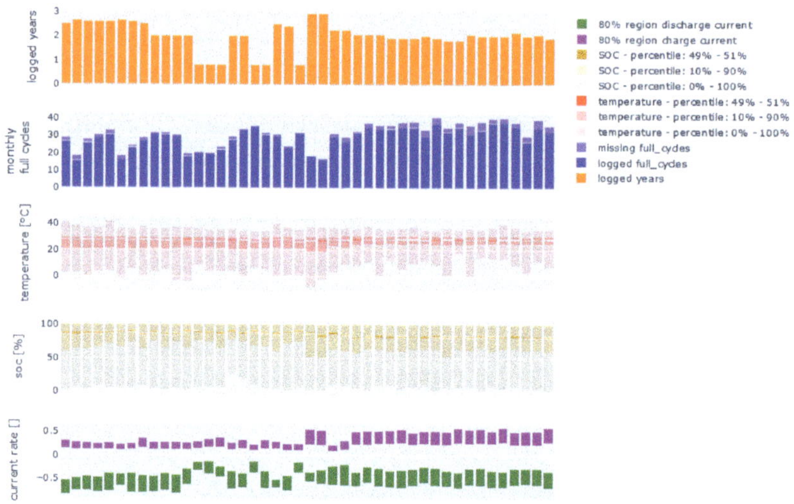

Figure 2. Usage profile of the buses of the fleet. Each entry represents one bus. The second top plot shows the mean monthly full cycles. The plot below indicates the distribution of temperature and SOC. The bottom plot shows the current range during operation as well as during charging.

2.2. Ageing Model Structure

The aim of the ageing model is the prediction of capacity fade (and resistance increase) under various operation conditions. State-of-the-art approaches include simple Ah counting up to physical modelling of ageing phenomena. An overview of degradation mechanisms and prediction algorithms is presented in [24].

Since the ageing behaviour depends strongly on the cell used and the operating conditions, the models must be parameterized on the basis of ageing tests. So far, only laboratory data have been used for this task. To accomplish this, a suitable model structure is necessary. An approach based on stress maps as proposed in [3] is used here. These types of models are also refereed to as weighted Ah models [25]. The resulting model structure is sketched in Figure 3.

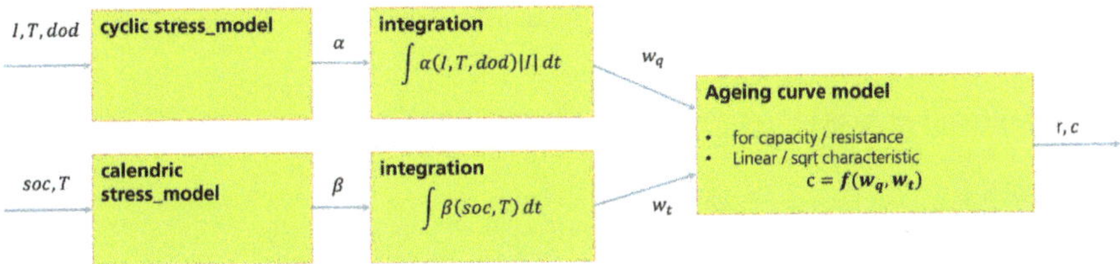

Figure 3. Structure of prediction model.

The ageing model consists of three submodels. The inputs and outputs are listed in Table 2. The parameters that can be trained depend on the approach used and are explained in more detail in the corresponding Sections 2.2.1 and 2.2.2.

Table 2. Inputs and outputs of the submodels.

Submodel	Inputs	Output	Trainable Parameter
calendric stress model	$x = [soc, T]$	$y = \beta$	weights Θ_β or parameter $p_9, \ldots, p_{13}$
cyclic stress model	$x = [T, dod, I]$	$y = \alpha$	weights Θ_α or parameter $p_0, \ldots, p_8$
ageing curve model	$x = [w_q, w_t]$	$y = c$	l_q and l_t

The main equations are as follows:

$$w_q(t) = \frac{1}{2C_{nom}} \int_{t_0}^{t} \alpha(I(t'), T(t'), dod(t'))|I(t')|dt' \quad (1)$$

$$w_t(t) = \frac{1}{S_y} \int_{t_0}^{t} \beta(soc(t'), T(t'))dt' \quad (2)$$

$$c(t) = f_{ac}(w_q(t), w_t(t)) = c_0 - (c_0 - c_{eol}) \cdot \left(\sqrt{\frac{w_q(t)}{l_q}} + \sqrt{\frac{w_t(t)}{l_t}} \right) \quad (3)$$

The value α describes the stress level of the operation during usage. It is used as a weighting factor for the charge throughput. The input dependencies are restricted to depth of discharge (DOD), temperature, and current. SOC is not used as an input due to complications in the parametrization process (see Section 2.3.1) and empirical observations. Analogue β indicates the calendric stress level. Integration of α and β over time using Equations (1)–(3) results in the weighted equivalent full cycles w_q and the weighted lifetime w_t, respectively. The factor $2C_{nom}$ is used for scaling to full cycles and the factor S_y is used for scaling to years.

The ageing curve f_{ac} is used to calculate the predicted capacity from these values. Here, a square-root ageing behaviour is assumed, as the checkup results (see Figure 1) indicate.

c_{eol} is the capacity at end of life, which is typically set to 80% of the initial capacity c_0. l_q is the cycle life in equivalent full cycles and l_t the calendric lifetime in years. The ageing curve depends heavily on the cell chemistry and must therefore be adapted. As explained in Section 2.3.3, transfer learning is used for this purpose.

The crucial point is the calculation of the cyclical and calendar stress factors. Various approaches are used in this thesis, including both a function-based model and a neural network. Details of these models are described in the following Sections 2.2.1 and 2.2.2. These two models are coupled using the approach described in Section 2.3.2. The resulting ageing model is referred to as a "coupled neural network" in this article.

In Section 3.3, a simple equal stress model is also used for benchmark purposes. Stress factors α and β are set to one regardless of operation conditions in this model. As the focus is on capacity estimation, the ageing models are limited to the prediction of capacity. However, they can be extended to predict resistance values.

2.2.1. Function-Based Stress Model

This model uses known functional relationships between the stress factors α and β and operation features (SOC, temperature, dod, and current rate). For temperature, the Arrhenius law is used. Depth of discharge as well as state of charge are modelled using a linear function. Piecewise linear interpolation is used for the current. The vector $I = (I_0, \ldots, I_4)$ is used to specify the grid points. I_0 is the maximal discharge current and I_4 the maximal charge current. Using these values, differences between charge (positive current) and discharge (negative current) as well as the absolute value of the current are taken into account.

$$\alpha_T(T) = p_0 + p_1 e^{p_2 T} \tag{4}$$

$$\alpha_{dod}(dod) = p_2 + p_3 \cdot dod \tag{5}$$

$$\alpha_I(I) = \begin{cases} p_4 + \frac{p_5 - p_4}{I_1 - I_0} \cdot (I - I_0) & I_0 < I \leq I_1 \\ \cdots & \\ p_7 + \frac{p_8 - p_7}{I_4 - I_3} \cdot (I - I_3) & I_3 < I \leq 4 \end{cases} \tag{6}$$

$$\beta_T(T) = p_9 + p_{10} e^{p_{11} T} \tag{7}$$

$$\beta_{soc}(soc) = p_{12} + p_{13} \cdot soc \tag{8}$$

The model contains a total of 14 parameters, which are combined to form the vector $p = (p_0, \ldots, p_{13})$. The resulting cyclic and calendric stress are obtained by multiplying the single stress factors (as proposed in [26]):

$$\begin{aligned} \alpha(T, dod, I) &= \alpha_T(T) \cdot \alpha_{dod}(dod) \cdot \alpha_I(I) \\ \beta(T, soc) &= \beta_T(T) \cdot \beta_{soc}(soc) \end{aligned} \tag{9}$$

2.2.2. Neural Network Stress Model

Two feedforward neural networks (FFNNs) are used to obtain calendric and cyclic stress factors from input features $x = [T, dod, I]$. The network structure is sketched in Figure 4.

A similar neural network is used to predict the calendric stress value β. This means functions (9) are replaced by

$$\begin{aligned} \alpha(T, dod, I) &= f_\alpha(T, dod, I; \Theta_\alpha) \\ \beta(T, soc) &= f_\beta(soc, T; \Theta_\beta) \end{aligned} \tag{10}$$

Here, Θ_α and Θ_β represent the weights of the two FFNNs. Since there are only a few input features and no exotic inter-dependencies, the network should be able to capture the information obtained from the given ageing test data. In order to guarantee positive stress values, activity regularization is used. Due to its superior mathematical properties, including differentiability, boundness, stationarity, and smoothness, the GELU activation function was used (see [27]). Information about thee number of nodes and further hyperparameters is given in Table 3.

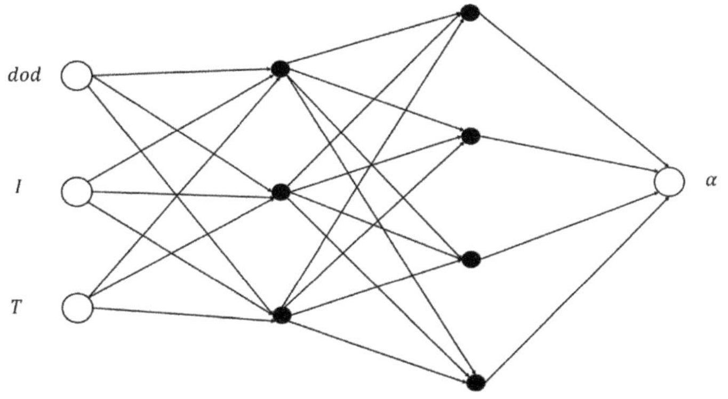

Figure 4. Structure of FFNN for cyclic stress value α. The number of nodes in each layer, the activation function, and further hyperparameters are listed in Table 3.

Table 3. Hyperparameters of the neural networks.

Parameter	Description	Value
n_{cal}	number of nodes in each layer of the calendric stress neural network	[5,5]
n_{cyc}	number of nodes in each layer of cyclic stress neural network	[8,8]
a	activation function	GELU
λ_{nonneg}	regularization constant for positive output values	1.0

2.3. *Parametrization of Ageing Models*

2.3.1. Using Laboratory Measurements

Since operation conditions are kept constant within the experiments, Equation (1) simplifies to

$$c(T,\overline{soc},dod,I_c,I_d,a_q,a_t) = c_0 - (c_0 - c_{eol}) \left(\sqrt{\frac{a_q}{l_q}} \cdot \sqrt{\frac{\alpha(I_c,T,dod) + \alpha(I_d,T,dod)}{2}} + \sqrt{\frac{a_t}{l_t}} \sqrt{\beta(\overline{soc},T)} \right) \tag{11}$$

$$a_t = \frac{t}{S_y} \tag{12}$$

$$a_q = \frac{t}{2C_{nom}} \int |I| dt = \frac{t}{2C_{nom}} \cdot \frac{I_c + |I_d|}{2} \tag{13}$$

Here, a_q denotes the (unweighted) equivalent full cycles and a_t the lifetime of the battery. It must also be emphasized that operation conditions are kept fixed; this is obviously not possible for SOC and current during cyclic experiments. This must be considered within the integration. Degradation effects during charging and discharging are superimposed. Hence, the charging current I_c as well as the discharge current I_d appear in the final Equation (11). SOC changes linearly with time in cyclic experiments. Nevertheless, analytical integration can be challenging if complex SOC dependencies are considered. In addition to empirical observations, this is one reason why the SOC is only taken into account for calendric ageing. Furthermore, a linear function is used (see Equation (8)). The solution of the integral therefore leads to the use of the mean SOC ($\overline{soc}$) in the calculation of the calendric stress factor.

These calculations and considerations lead to a fairly simple equation for predicting the capacity from ageing test experiments. All integrations are solved analytically, which is a great benefit for subsequent parameter optimization. Nevertheless, the equation is valid for cyclic and calendar tests, allowing both stress models to be parameterized together. There is no need for a two-stage fitting approach (first calendar ageing and then cyclic ageing), which is commonly performed.

From Equation (11), it can be seen that there is a degree of freedom in determining the calendar and cycle life parameters l_q and l_t (multiplying l_t as well as β with same factor will not change the predicted capacity). This is used to specify reference conditions for which $\beta(soc_r, T_r) = 1$ and $\alpha(I_r, T_r, dod_r) = 1$ hold. The used reference conditions are listed in Table 4.

Table 4. Reference conditions.

Parameter	Value
soc_r	0.8
T_r	25 °C
dod	0.3
I_r	−0.5

2.3.2. Coupling of Function-Based Model and Neural Network

Neural networks or even lookup tables have great potential in accurately predicting stress map functions α and β. However, a large number of experiments/cells would be needed to ensure good extrapolation behaviour.

With physics-informed neural networks, a similar problem is tackled [28]. Neural networks are combined with physical equations to gain good extrapolation properties but still preserve good fit capability to data. Therefore, the presented functional stress model is coupled with the feedforward neural networks presented in Section 2.2.2. Instead of only considering the deviation of J_{data} from training data, a second term is added to the loss function J. This term describes the deviation between the two models at specified collocation points. These collocation points span a dense grid of conditions where neural networks will predict valid results. In summary, the following four vectors are involved in the calculation of the cost function:

- $c_{t,m}$: capacities obtained during checkup tests;
- $c_{t,nn}$: capacities predicted by neural networks under experiment conditions;
- $c_{col,em}$: capacities predicted by an empirical model for collocation points;
- $c_{col,nn}$: capacities predicted by neural networks for collocation points.

And the used loss function reads as follows:

$$J(c_{t,m}, c_{t,nn}, c_{col,em}, c_{col,nn}) = J_{data}(c_{t,m}, c_{t,em}) + \lambda_{pinn} \cdot J_{col}(c_{col,em}, c_{col,nn}) \quad (14)$$

The second term J_{col} describes the deviation of the neural network from the functional model output at collocation points $c_{col,em}$. Through minimizing this loss function, a good fit to training data as well as good extrapolation behaviour is obtained. J_{col} basically serves as a regularization term that couples the functional model with the neural network and prevents overfitting.

As usual, the mean squared error is used for both loss functions. For J_{data}, the sum of overall training points is taken; for J_{col}, collocation points are used.

The Tensorflow framework is used to create and train the ageing model, including the collocation loss term. The established Adam algorithm (see [29]) is used for optimization.

2.3.3. Transfer Learning

Transfer learning (TL) is used if a model exists for a general task, but needs to be adapted to a specific application. Here, it is assumed that the stress level is the same for different cells. However, both the cycle and calender lifetimes as well as the ageing characteristics (linear or quadratic ageing) differ considerably between different cell chemistries. To take this into account, the ageing curve of the model is fitted to the cell under consideration. If laboratory ageing tests are available, the same procedure as described in Section 2.3.1 (but with fixed stress map weights) can be used.

The situation is somewhat different when field data are used for this purpose. Here, the operating conditions such as temperature and current are not kept constant, but change rapidly. Hence, the integral in Equation (1) does not vanish and must be calculated using numerical approaches. To speed this up, histogram-based methods (see [30]) are used. As a result, weighted full cycles $w_{q,i}$ and weighted lifetime values $w_{t,i}$ are obtained at each checkup position. They can be used as input values to fit the ageing curve parameters l_q and l_t by minimizing

$$\sum_{i}^{N}\left(f_{ac}(w_{q,i}, w_{t,i}; l_q, l_t) - c_i\right)^2 \to \min \tag{15}$$

Here, c_i are the estimated capacities from the checkup tests.

3. Results

3.1. Estimated Stressmaps

In this section, the resulting stress maps for the presented model structures are discussed. The Figures 5–7 show the fit results for the calendric model part. It can be seen that the function model is suitable for extrapolation, whereas the neural network gives unreasonable values at conditions not covered by ageing tests.

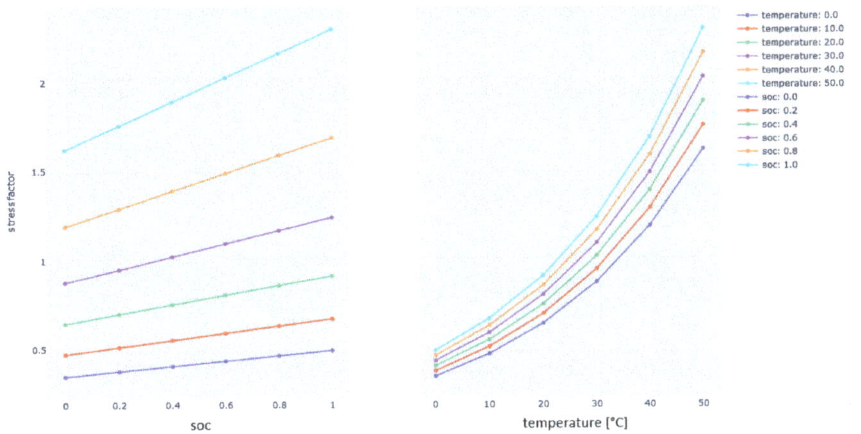

Figure 5. Calendric stress map obtained from function-based model. The values are sensible and allow good extrapolation over the whole temperature and SOC range.

Figure 8 compares the prediction results of the three models. As expected, the neural network produces the lowest prediction error. On the other hand, extrapolation performance for unobserved data is poor. One indicator for this is the smoothness of the stress map, defined as the inverse of the integral of the hessian (see [31]):

$$S(\beta) = \frac{1}{\int \frac{\partial\beta(soc,T)}{\partial soc \partial T} dsocdT} \tag{16}$$

Low values of this metric indicate a rugged stress map and potential overfitting. Thus, extrapolation performance will also be poor in general. A summary of RMSE as well as smoothness is given in Table 5. The coupled neural network is a compromise between the function-based model and the neural network: it offers good extrapolation properties and good prediction accuracy. For the cyclic part, the deviation between the function model and coupled neural network is larger (see Appendix A). That indicates that the coupled approach significantly increases the prediction accuracy compared to the functional model.

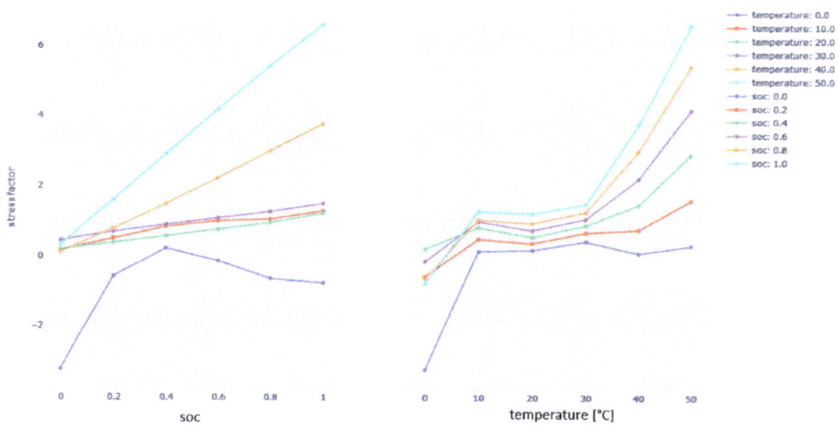

Figure 6. Calendric stress map obtained from neural network. Main trends and relationships are as expected. However, for some operating conditions (e.g., at a temperature of 0 °C) for which no experiments were carried out, the values are not meaningful. Therefore, the stress map is not suitable for extrapolation to conditions that are not included in the training dataset.

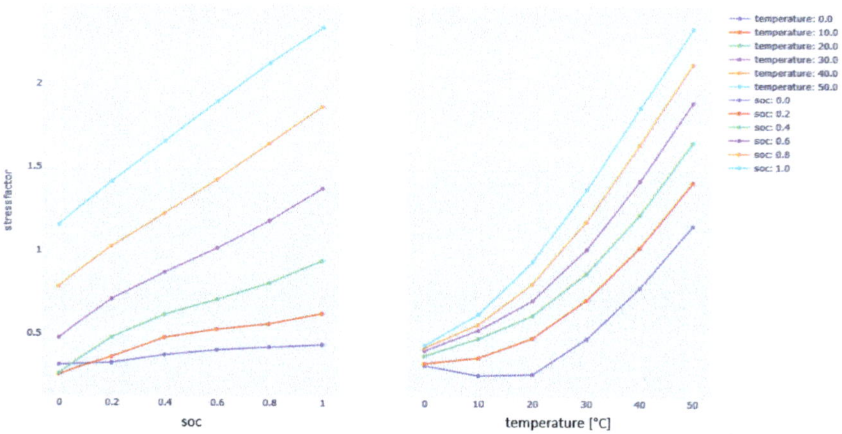

Figure 7. Calendric stressmap obtained from coupled model. There are only slight deviations from the function-based model. This means that this model is already capable of modelling calendric ageing with good accuracy.

Table 5. Capacity prediction error for training data and smoothness metric for stress maps.

Model	RMSE	Max Error	Stressmap Smoothness
function-based model	5.4%	19.1%	9.5
neural network	2.3%	11.1%	0.3
coupled neural network	4.5%	16.9%	0.9

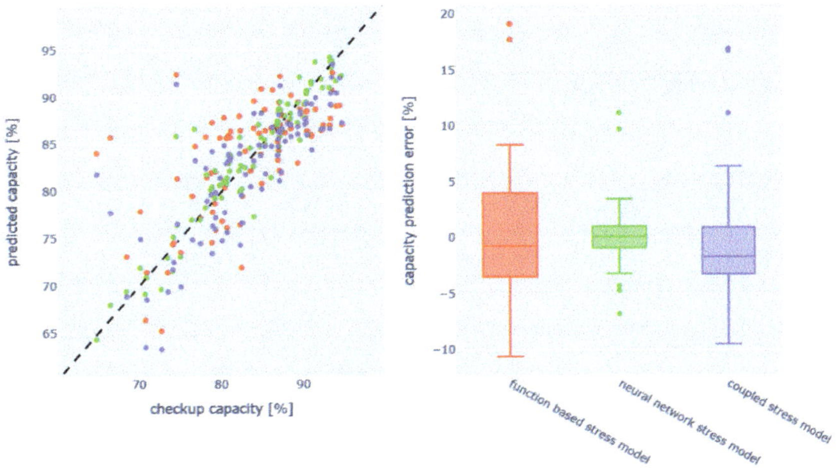

Figure 8. **(left)** Prediction results of the three models for the used training dataset plotted against the target checkup test capacities. **(right)** Box plot of prediction error.

3.2. Application to Other Cell Chemistries

In order to validate the generalization properties of the estimated stress map, it was applied to the laboratory ageing test campaigns listed in Table 1. Here, dedicated capacity checkup tests were performed at regular intervals.

In the first step, the ageing model was parameterized using the Sony dataset and applied to the batteries 2020 and the Kokam ageing test campaigns (see Section 2.1). As described in Section 2.3.3, the cycle life and calendar life parameters of the ageing curve model were adjusted before predicting the capacity. The result is shown in Figure 9. The standard deviation of the prediction error (summarized in Table 6) is around 5% for both ageing test campaigns and within the same magnitude as for the training dataset. This confirms the analysis performed in [32], which states that general ageing effects can be transferred between cell chemistries.

Table 6. Capacity prediction error for investigated measurement campaigns.

Campaign	RMSE	Max Error
Sony (training dataset)	4.5%	16.9%
Kokam	5.3%	14.3%
batteries 2020	5.0%	10.8%

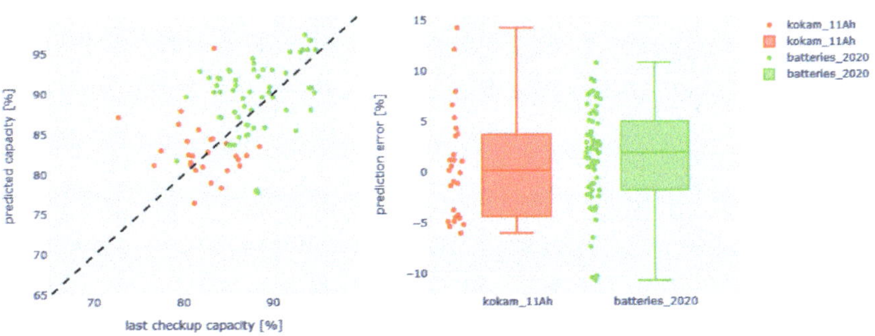

Figure 9. Plot of determined capacity at checkups compared to the prediction of the coupled neural network. Prediction error is slightly higher than on training dataset (see Figure 8).

3.3. Application to Electric Vehicle Fleet

The capacities for the field data checkups were predicted as outlined in Section 2.3.3. As a benchmark, two further prediction methods were applied. In summary, the following algorithms are compared:

- Equal stress model: instead of using stress maps to evaluate the operation conditions, only the pure lifetime a_t and equivalent full cycles a_q are used.
- Coupled neural network without transfer learning: the coupled neural network without adjustment to field data checkup tests is used. This means that the transfer learning step is skipped.
- Coupled neural network with transfer learning: the coupled neural network with adjustment to field data checkup tests is used.

The results are shown in Figure 10 and summarized in Table 7. It can be seen that the coupled neural network is superior to the equal stress model. The use of transfer learning to adapt to the different cell chemistries improves the estimation. Whereas the other models overestimate the capacity fade, the prediction errors are distributed around zero for this approach.

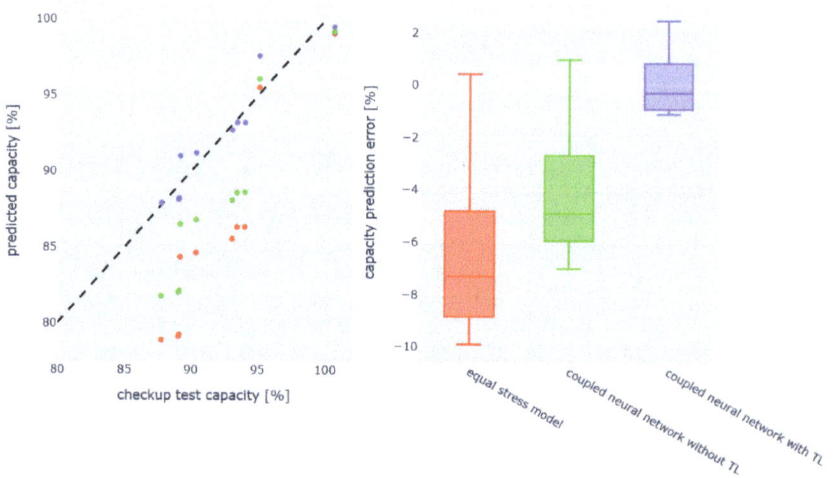

Figure 10. Plot of measured capacity at checkups compared to the predicted capacity using three different approaches. Field-data-tuned stress map provides lowest prediction error.

Table 7. Capacity prediction error for electric vehicle fleet.

Campaign	RMSE	Max Error
equal stress model	7.1%	9.9%
coupled neural network without TL	4.8%	7.0%
coupled neural network with TL	1.2%	2.5%

4. Discussion

Previous publications have investigated the transferability of ageing models between cell chemistries on the basis of laboratory data. This study extends this approach, as it shows how laboratory ageing test data may be combined with field data in order to improve capacity prediction and thereby the end-of-life prognosis of Li-ion batteries. Validation was enabled by conducting reference tests on multiple vehicles of an electric fleet.

The usage of neural networks coupled to function-based models was applied to create accurate but also reliable stress maps. In particular, extrapolation outside the–often sparse–ageing test data is ensured. Transfer learning techniques were used to adapt cell-specific ageing characteristics. The transfer of general, extensively studied ageing effects from laboratory measurements to field data have proven to be feasible.

Future work should extend the usage of field data. For instance, SOH estimations could be used instead of time- and cost-intensive checkup tests.

Author Contributions: Conceptualization, T.L.; methodology, T.L. and E.B.; software, T.L.; validation, T.L. and E.B.; formal analysis, T.L.; investigation, T.L.; resources, E.B. and R.K.; data curation, R.K.; writing—original draft preparation, T.L.; writing—review and editing, T.L., E.B. and G.S.; visualization, T.L.; supervision, T.L. and R.K.; project administration, R.K.; funding acquisition, R.K. All authors have read and agreed to the published version of the manuscript.

Funding: We gratefully acknowledge the financial support by Bundesministerium für Bildung und Forschung (BMBF 03XP0308A).

Data Availability Statement: The data presented in this study are available on request from the corresponding author on reasonable request. The data are not publicly available due to privacy policies.

Conflicts of Interest: The authors declare no conflicts of interest.

Abbreviations

The following abbreviations are used in this manuscript:

SOC	State of Charge
DOD	Depth of Discharge
SOH	State of Health
FFNN	Feedforward Neural Network
TL	Transfer Learning
RMSE	Root Mean Squared Error
IVI	Institute for Transportation and Infrastructure Systems

Appendix A

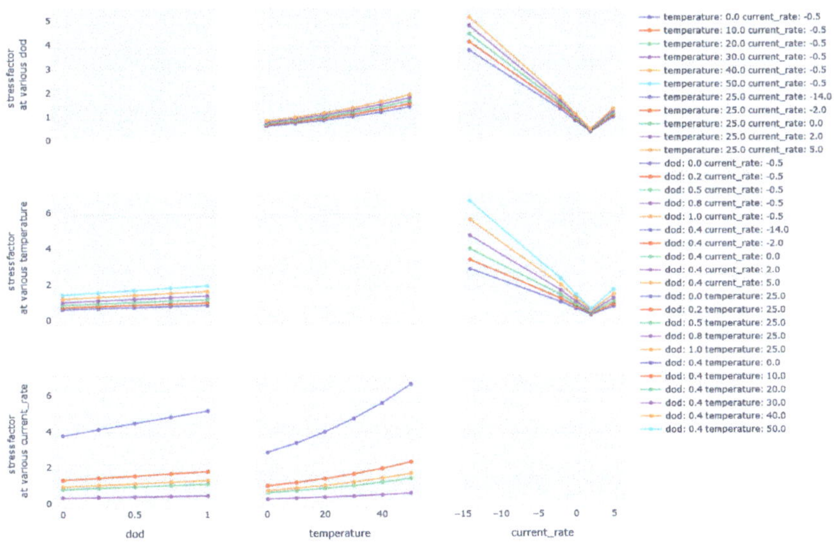

Figure A1. Cyclic stress map obtained from function-based model.

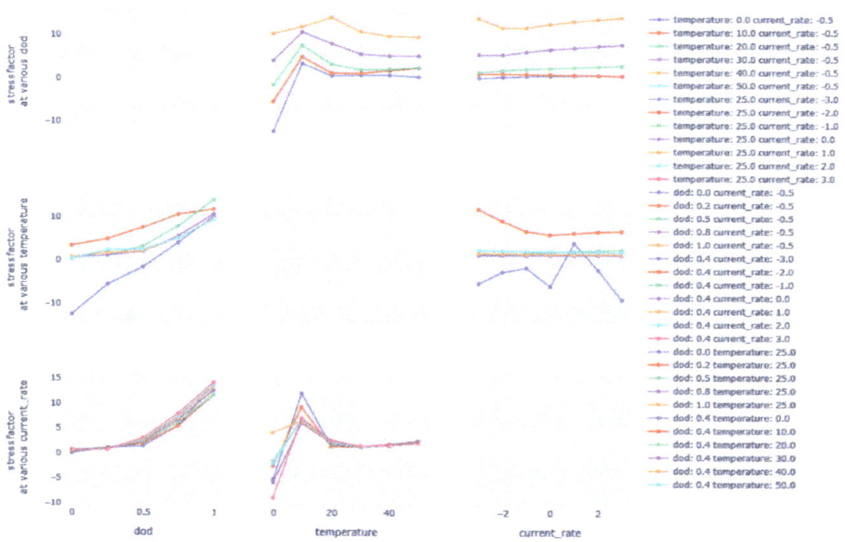

Figure A2. Cyclic stress map obtained from neural network.

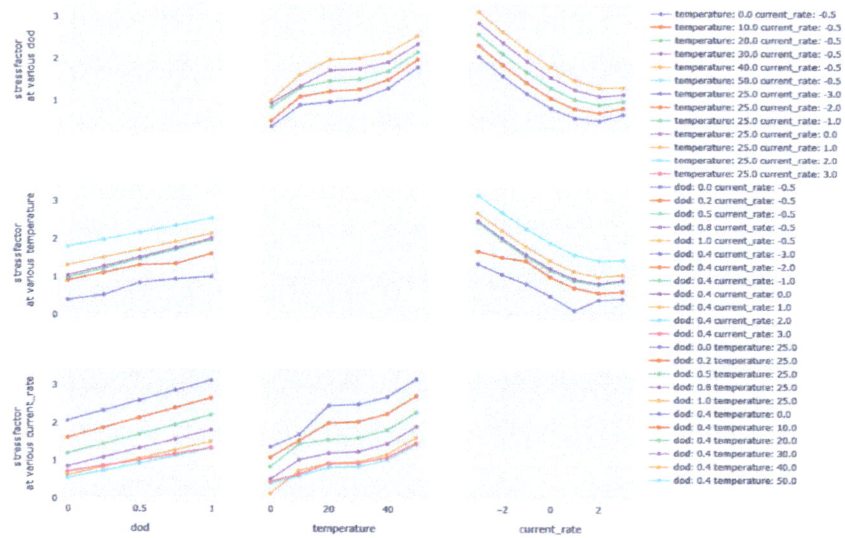

Figure A3. Cyclic stress map obtained from coupled model.

References

1. Liao, Z.; Lv, D.; Hu, Q.; Zhang, X. Review on Aging Risk Assessment and Life Prediction Technology of Lithium Energy Storage Batteries. *Energies* **2024**, *17*, 3668. [CrossRef]
2. Gewald, T.; Candussio, A.; Wildfeuer, L.; Lehmkuhl, D.; Hahn, A.; Lienkamp, M. Accelerated Aging Characterization of Lithium-ion Cells: Using Sensitivity Analysis to Identify the Stress Factors Relevant to Cyclic Aging. *Batteries* **2020**, *6*, 6. [CrossRef]
3. Vermeer, W.; Member, S.; Mouli, G.R.C.; Bauer, P. A Comprehensive Review on the Characteristics and Modeling of Lithium-Ion Battery Aging. *IEEE Trans. Transp. Electrif.* **2022**, *8*, 2205–2232. [CrossRef]
4. Heimhuber, P. Parametrisierung und Evaluierung von Alterungsmodellen für NMC- und NCA- basierte Lithium-Ionen-Batterien. Diploma Thesis, TU Dresden, Dresden, Germany, 2022.

5. Birkl, C.; Roberts, M.R.; McTurk, E.; Bruce, P.G.; Howey, D.A. Degradation diagnostics for lithium ion cells. *J. Power Sources* **2017**, *341*, 373–386. [CrossRef]
6. Schmitt, J.; Schindler, M.; Oberbauer, A.; Jossen, A. Determination of degradation modes of lithium-ion batteries considering aging-induced changes in the half-cell open-circuit potential curve of silicon–graphite. *J. Power Sources* **2022**, *532*, 231296. [CrossRef]
7. Xu, R.; Wang, Y.; Chen, Z. Data-Driven Battery Aging Mechanism Analysis and Degradation Pathway Prediction. *Batteries* **2023**, *9*, 129. [CrossRef]
8. Smith, A.J.; Svens, P.; Varini, M.; Lindbergh, G.; Lindström, R.W. Expanded In Situ Aging Indicators for Lithium-Ion Batteries with a Blended NMC-LMO Electrode Cycled at Sub-Ambient Temperature. *J. Electrochem. Soc.* **2021**, *168*, 110530. [CrossRef]
9. Che, Y.; Zheng, Y.; Forest, F.E.; Sui, X.; Hu, X.; Teodorescu, R. Predictive health assessment for lithium-ion batteries with probabilistic degradation prediction and accelerating aging detection. *Reliab. Eng. Syst. Saf.* **2024**, *241*, 109603. [CrossRef]
10. Xu, X.; Tang, S.; Han, X.; Lu, L.; Wu, Y.; Yu, C.; Sun, X.; Xie, J.; Feng, X.; Ouyang, M. Fast capacity prediction of lithium-ion batteries using aging mechanism-informed bidirectional long short-term memory network. *Reliab. Eng. Syst. Saf.* **2023**, *234*, 109185. [CrossRef]
11. Román-Ramírez, L.; Marco, J. Design of experiments applied to lithium-ion batteries: A literature review. *Appl. Energy* **2022**, *320*, 119305. [CrossRef]
12. Mayemba, Q.; Mingant, R.; Li, A.; Ducret, G.; Venet, P. Aging datasets of commercial lithium-ion batteries: A review. *J. Energy Storage* **2024**, *83*, 110560. [CrossRef]
13. Sulzer, V.; Mohtat, P.; Aitio, A.; Lee, S.; Yeh, Y.T.; Steinbacher, F.; Khan, M.U.; Lee, J.W.; Siegel, J.B.; Stefanopoulou, A.G.; et al. The challenge and opportunity of battery lifetime prediction from field data. *Joule* **2021**, *5*, 1934–1955. [CrossRef]
14. von Bulow, F.; Meisen, T. A review on methods for state of health forecasting of lithium-ion batteries applicable in real-world operational conditions. *J. Energy Storage* **2023**, *57*, 105978. [CrossRef]
15. Liu, K.; Peng, Q.; Che, Y.; Zheng, Y.; Li, K.; Teodorescu, R.;Widanage, D.; Barai, A. Transfer learning for battery smarter state estimation and ageing prognostics: Recent progress, challenges, and prospects. *Adv. Appl. Energy* **2023**, *9*, 100117. [CrossRef]
16. Azkue, M.; Lucu, M.; Martinez-Laserna, E.; Aizpuru, I. Calendar Ageing Model for Li-Ion Batteries Using Transfer Learning Methods. *World Electr. Veh. J.* **2021**, *12*, 145. [CrossRef]
17. Zhou, K.Q.; Qin, Y.; Yuen, C. Transfer-Learning-Based State-of-Health Estimation for Lithium-Ion BatteryWith Cycle Synchronization. *IEEE/ASME Trans. Mechatron.* **2022**, *28*, 692–702. [CrossRef]
18. von Bulow, F.; Mentz, J.; Meisen, T. State of health forecasting of Lithium-ion batteries applicable in real-world operational conditions. *J. Energy Storage* **2021**, *44*, 103439. [CrossRef]
19. Li, S.; He, H.; Zhao, P.; Cheng, S. Health-Conscious vehicle battery state estimation based on deep transfer learning. *Appl. Energy* **2022**, *316*, 119120. [CrossRef]
20. BMBF. Battnutzung-Cluster. 2022. Available online: https://www.battnutzung-cluster.de/de/projekte/febal/ (accessed on 29 March 2023).
21. Krupp, A. Semi-Empirical Aging Model for Predicting the Capacity Loss of Lithium-Ion Batteries in Stationary Storage Systems. Doctoral Dissertation, Carl von Ossietzky Universität Oldenburg, Oldenburg, Germany, 2023.
22. Wildfeuer, L.; Karger, A.; Aygül, D.;Wassiliadis, N.; Jossen, A.; Lienkamp, M. Experimental degradation study of a commercial lithium-ion battery. *J. Power Sources* **2023**, *560*, 232498. [CrossRef]
23. Timmermans, J.M.; Nikolian, A.; De Hoog, J.; Gopalakrishnan, R.; Goutam, S.; Omar, N.; Coosemans, T.; Van Mierlo, J.; Warnecke, A.; Sauer, D.U.; et al. Batteries 2020—Lithium-ion battery first and second life ageing, validated battery models, lifetime modelling and ageing assessment of thermal parameters. In Proceedings of the 2016 18th European Conference on Power Electronics and Applications (EPE'16 ECCE Europe), Karlsruhe, Germany, 5–9 September 2016; pp. 1–23. [CrossRef]
24. Hu, X.; Xu, L.; Lin, X.; Pecht, M. Battery Lifetime Prognostics. *Joule* **2020**, *4*, 310–346. [CrossRef]
25. Redondo-Iglesias, E.; Venet, P.; Pelissier, S. Modelling Lithium-Ion Battery Ageing in Electric Vehicle Applications—Calendar and Cycling Ageing Combination Effects. *Batteries* **2020**, *6*, 14. [CrossRef]
26. Xu, B.; Oudalov, A.; Ulbig, A.; Andersson, G.; Kirschen, D.S. Modeling of Lithium-Ion Battery Degradation for Cell Life Assessment. *IEEE Trans. Smart Grid* **2018**, *9*, 1131–1140. [CrossRef]
27. Lee, M. Mathematical Analysis and Performance Evaluation of the GELU Activation Function in Deep Learning. *J. Math.* **2023**, *1*, 4229924. [CrossRef]
28. Cuomo, S.; Cola, V.D.; Giampaolo, F.; Rozza, G.; Raissi, M.; Piccialli, F. Scientific Machine Learning Through Physics–Informed Neural Networks: Where we are and What's Next. *J. Sci. Comput.* **2022**, *92*, 88. [CrossRef]
29. Kingma, D.P.; Ba, J.L. Adam: A method for stochastic optimization. *arXiv* **2017**, arXiv:1412.6980.
30. Zdravevski, E.; Lameski, P.; Mingov, R.; Kulakov, A.; Gjorgjevikj, D. Robust histogram-based feature engineering of time series data. In Proceedings of the 2015 Federated Conference on Computer Science and Information Systems (FedCSIS), Lodz, Poland, 13–16 September 2015; pp. 381–388. [CrossRef]

31. Chen, X.; Tong, Z.; Liu, H.; Cai, D. Metric learning with two-dimensional smoothness for visual analysis. In Proceedings of the 2012 IEEE Conference on Computer Vision and Pattern Recognition 2012, Providence, RI, USA, 16–21 June 2012; pp. 2533–2538. [CrossRef]
32. Deletang, T.; Barnel, N.; Franger, S.; Assaud, L. Transposition of a weighted ah-throughput model to another li-ion technology: Is the model still valid? New insights on the mechanisms. In Proceedings of the International Conference on Computational Methods for Coupled Problems in Science and Engineering, Rhodes Island, Greece, 12–14 June 2017.

batteries

MDPI

Article

State of Health Estimation for Lithium-Ion Batteries Based on Transition Frequency's Impedance and Other Impedance Features with Correlation Analysis

Mohammad K. Al-Smadi and Jaber A. Abu Qahouq *

Department of Electrical and Computer Engineering, College of Engineering, The University of Alabama, Tuscaloosa, AL 35487, USA; mkalsmadi@crimson.ua.edu
* Correspondence: jaberq@eng.ua.edu; Tel.: +1-205-348-8669

Abstract: This paper presents data-driven impedance-based state of health (SOH) estimation for commercial lithium-ion batteries across an SOH range of ~96% to ~60%. Battery health indicators at the transition frequency of the battery impedance Nyquist plot are utilized to develop an SOH estimator based on an artificial neural network (ANN). In addition, two more ANN-based SOH estimators utilizing some impedance magnitude and phase values are developed. Spearman correlation analysis is utilized to identify the frequency points at which the impedance magnitude and phase values show strong correlations with SOH values and are thus utilized as SOH indicators. The performance evaluation of the developed SOH estimators shows that the maximum root mean square error (RMSE) is equal to 1.39%, the maximum mean absolute error (MAE) is equal to 1.25%, the maximum mean absolute percentage error (MAPE) is equal to 1.55%, and the minimum coefficient of determination (R^2) is equal to 0.983.

Keywords: artificial neural network; battery; correlation analysis; impedance; state of health estimation; machine learning; electrochemical impedance spectroscopy; magnitude; phase; Nyquist; transition frequency

1. Introduction

Lithium-ion batteries have been widely adopted across a range of applications, from electric vehicles (EVs) to consumer electronics, owing to several advantages such as high energy density and low internal discharge [1]. To meet the increasing performance requirements of battery packs, developing advanced battery management systems (BMS) has become an essential step. One of the most important BMS functions is the state of health (SOH) estimation [2–6]. Battery SOH estimation is crucial in prolonging the lifetime of battery systems and maintaining their safety. SOH estimation is utilized in monitoring the health of the battery, suggesting better battery charging and discharging strategies, and triggering protection functions to avoid battery failures. Recently, SOH estimation gained more attention in second-use (i.e., retired) battery applications. In such applications, SOH estimation is essential for the safe and efficient utilization of second-use batteries [7–9].

Battery aging (i.e., health deterioration) causes include, but are not limited to, high/low temperature outside normal range, over/under voltage, and mechanical stress [10]. Battery aging types can be classified into calendar aging and cyclic aging [10]. Calendar aging is due to side reactions while the battery is not in use, whereas cyclic aging is due to repeated battery charging and discharging (cycling). Several mechanisms govern battery aging, such as solid electrolyte interface (SEI) formation and thickening, cathode electrolyte interface

(CEI) formation, lithium plating, and binder decomposition [11]. Generally, battery health degradation appears as power fade and capacity fade. In terms of capacity fade, the SOH value for a battery can be defined as the ratio between the available capacity $Q_{\text{available}}$ and the nominal capacity Q_{nominal} as expressed in (1):

$$SOH = Q_{\text{available}}/Q_{\text{nominal}} \quad (1)$$

where the available capacity $Q_{\text{available}}$ is the amount of charge that a fully charged battery can provide when the battery is discharged to its minimum voltage, where Q_{nominal} is provided by the manufacturer.

Many SOH estimation techniques were proposed in the literature with various classifications [5,6,9,12–17]. The classifications are based on different factors such as the variables needed to estimate the SOH, the need for direct measurement, and whether circuit and/or physical models are utilized or not. Generally, SOH estimation techniques can be classified into experimental techniques and model-based techniques, as illustrated in Figure 1. The white boxes at the bottom of Figure 1 include examples for each SOH estimation category/subcategory.

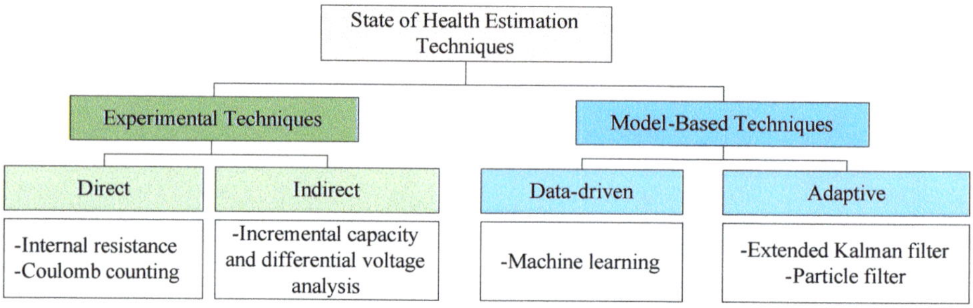

Figure 1. SOH estimation techniques.

Experimental techniques can be direct or indirect. Examples of direct techniques include utilizing the internal resistance and Coulomb counting to estimate the SOH. The internal resistance is affected by battery aging in such a way that its value increases as the battery ages (i.e., as the SOH value decreases) [18]. The current pulse method [19,20] is usually utilized to estimate the internal resistance. However, the internal resistance of a battery is a function of various factors such as temperature and state of charge (SOC) [21,22]. The Coulomb counting method is based on counting the Ampere-Hour (Ah) during battery discharging and charging. The Ah amount supplied/accepted by the battery during discharging/charging changes as the battery ages [3]. Despite its simplicity, the Coulomb counting method suffers from sensing error accumulation [23].

Incremental capacity (i.e., the derivative of the accumulated capacity during a charge/discharge process with respect to voltage, referred to as "IC") and differential voltage (dV) analyses utilize the relationship between the incremental capacity and the battery voltage to indirectly identify aging patterns and estimate the SOH value; for example, by utilizing the peaks in the voltage–IC curves [24,25].

Model-based techniques utilize various models to estimate the SOH value of a battery, and they are classified into adaptive and data-driven techniques. Adaptive SOH techniques are capable of adjusting and updating their parameters based on data obtained from real-time operation. Some examples of adaptive techniques include the Kalman filter [26–28] and particle filter [29–31].

Data-driven SOH estimation techniques have gained more interest over the past decade, especially with the advancements in computation capabilities. Without a need to know the underlying aging mechanisms, data-driven techniques capture aging patterns and extract features from battery historical data [3,32,33].

SOH estimation based on the battery impedance has drawn attention in the literature due to its speed and ability to reflect the battery's internal electrochemical processes [33,34]. Generally, SOH features are extracted from the impedance curves (the impedance magnitude, phase, and Nyquist). Utilizing battery impedance for SOH estimation might fall into one of the following three categories:

(1) Utilizing a broadband EIS curve to estimate the SOH value. The authors in [35] utilized the entire EIS curve of 60 frequency points (120 points of 60 real impedance values and 60 imaginary impedance values) to predict the SOH. However, this study was for 45 mAh coin cell batteries. Also, utilizing a high-dimensional input increases the computational cost.

(2) Identifying specific points from the impedance curves to be utilized as SOH features:

  a. Fixed-frequency impedance points: impedance values at specific frequencies are identified to have a strong correlation with the SOH value, such as the impedance magnitude at 1 kHz [36] and 316 Hz [37]; impedance real part at 0.1 Hz [38]; and the impedance imaginary part at 17.8 Hz and 2.16 Hz [35].

  b. Non-fixed frequency points: features are identified from the pattern at which the impedance curves change at different SOH values, such as the minimum impedance magnitude [39], the impedance magnitude when the impedance phase is equal to 0° [33], and the value of the frequency at which the impedance phase is equal to 0° [40].

  c. Features from the Nyquist plot (the plot of the impedance real part against the impedance imaginary part): By studying how Nyquist plots change with SOH, various features can be extracted to be utilized for SOH estimation. For example, at the Nyquist plot's peak point, the impedance real part, the impedance imaginary part, and the impedance vector amplitude were utilized for SOH estimation in [21].

  d. Features from the phase–magnitude plot: Another way to look at the battery's impedance spectrum is to plot the phase against the magnitude [34]. From the phase–magnitude plot, some SOH features can be obtained, such as the differential impedance magnitude, which is the difference in the impedance magnitude at the valley and peak of the phase–magnitude plot [34].

(3) Fitting the EIS curve to obtain values for the equivalent circuit model (ECM) parameters. For example, the EIS curve can be fit to obtain values for solid electrolyte interphase (SEI) resistance [41,42], diffusion constant phase elements [43], and charge transfer resistance [44]. The change in the ECM parameters is utilized for SOH estimation.

This paper contributes to different categories of those mentioned above by presenting different SOH estimators. First, an SOH estimator is developed based on the battery's real and imaginary impedance parts at the transition frequency within the impedance Nyquist plot (contributing to category 2.c from the categories above). Second, this paper contributes to category 2.a from the categories above by presenting two more SOH estimators: one utilizing features extracted from the impedance magnitude curve, and another utilizing features extracted from the impedance phase curve (that is not common in the literature). A correlation analysis is carried out to identify the frequency points at which the impedance magnitude and phase are utilized as indicators for SOH estimation. The performance of the developed SOH estimators is evaluated along with some insights. The batteries used in this paper were deeply aged to a highly deteriorated SOH value (~60%), which fills a gap in the literature for EIS data of batteries under deep degradation levels. The remainder of this paper is organized as follows: Section 2 introduces the battery aging and data collection protocol followed to obtain aging data needed for SOH estimation model training and testing. Section 3 presents the proposed SOH indicators for each estimator. Section 4 presents a correlation analysis to identify the key frequency range (or points) for impedance points (magnitude and phase) to be used in SOH estimators. Section 5 presents the SOH estimation model and its accuracy evaluation. Finally, the conclusions are drawn in Section 6.

2. Battery Aging and Data Collection Protocol

Three commercial LG INR18650 3500 mAh battery cells (by LG Chem Ltd., Seoul, Republic of Korea) [45] were cycled/aged in the laboratory according to the aging protocol shown in Figure 2.

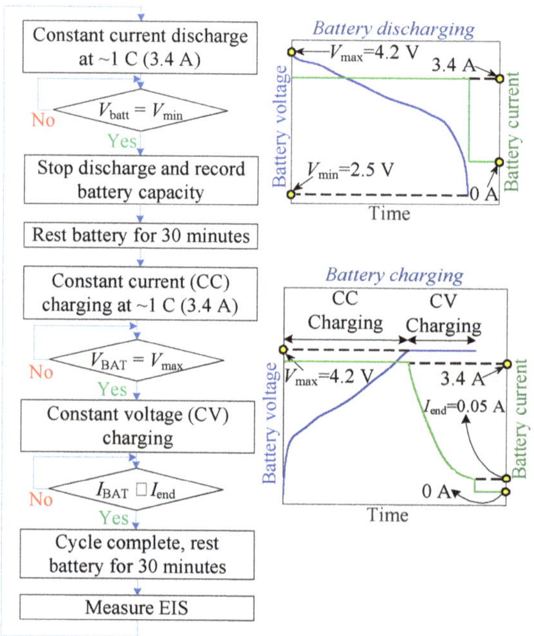

Figure 2. Battery aging and EIS data collection flowchart.

The main manufacturer's specifications for the cycled battery cells are listed in Table 1 [45]. Before cycling the battery cells, the initial capacity of each battery cell is calibrated using the Coulomb counting method by fully discharging the battery. Also, the EIS curve is measured for each battery cell to obtain the SOH features at the initial capacity level (the initial SOH value). Then, each battery cell undergoes a cycling process in which the battery cell is repeatedly discharged and charged. The battery cell is discharged at ~1 C (3.4 A). The battery cell discharging operation is utilized to measure the battery capacity at each cycle using the Coulomb counting method by integrating the battery discharge current over time during the discharging operation in each cycle. Battery cell charging starts with constant current (CC) charging mode in which the charging current is regulated at ~1 C (3.4 A). Once the battery cell voltage reaches the maximum value (i.e.,

4.2 V for the battery cells used in this paper), the constant voltage (CV) charging mode starts by regulating the battery voltage at 4.2 V until the battery current drops below the charging end current I_{end} (0.05 A for the battery cells used in this paper). The battery cell is rested for 30 min after each discharge operation and after each charge operation. The battery cells were cycled at room temperature (~23 °C) with a relative humidity in the range of 40–60%. Battery cell cycling and EIS measurements were carried out using the Squidstat Cycler—Base Model from Admiral Instruments (main specifications include 24-bit resolution Analog-to-Digital Converter "ADC", 320 nV minimum voltage resolution, 69.75 nA minimum current resolution, $\leq$0.5% Ω impedance magnitude accuracy, and $\leq$0.1° impedance phase offset [46]. The impedance spectrum curve is measured after every cycle. After the battery is fully charged, the battery is rested for 30 min before taking the EIS measurement. Several studies in the literature discussed the effect of resting time before taking EIS measurements. Essentially, the resting time should be selected such that the battery is allowed to recover and "relax" after the charging/discharging operation in order to obtain valid and consistent EIS measurements [47–49]. The frequency range for the EIS measurement is 10 kHz to 0.01 Hz with 10 points per decade, which results in 61 frequency points in total. The data collected during battery aging are used to train and test the artificial neural network (ANN)-based SOH estimators (presented in Section 5). Figure 3 shows the SOH value of each battery versus the number of cycles.

Table 1. Main manufacturer's specifications for LG INR18650MJ1-3.5Ah battery cell [45].

Nominal capacity C	3.5 Ah
Nominal voltage V_{nom}	3.635 V
Minimum voltage V_{min}	2.50 V
Maximum voltage V_{max}	4.2 V
Maximum charging current	3.4 A
Charing end current I_{end}	0.05 A

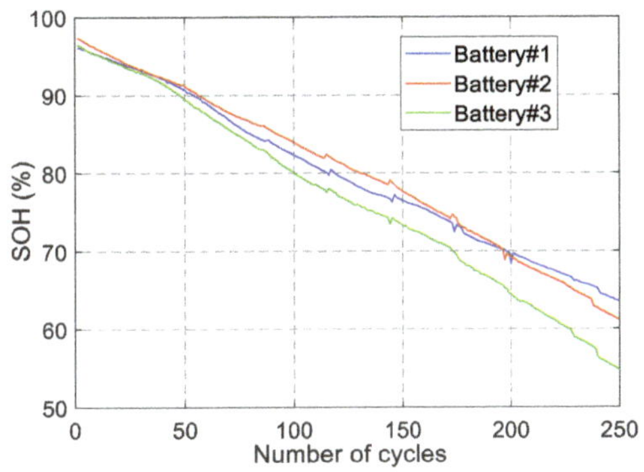

Figure 3. SOH vs. number of cycles.

3. SOH Indicators from the Electrochemical Impedance Spectrum

The three battery cells under study in this paper are referred to as BAT#1, BAT#2, and BAT#3. Sample EIS curves of BAT#1 at different aging cycles are shown in Figure 4. In this paper, multiple combinations of indicators obtained from the EIS curves are utilized to develop three SOH estimators ($SOH_{EST\#1}$, $SOH_{EST\#2}$, and $SOH_{EST\#3}$). The SOH estimators are explained as follows:

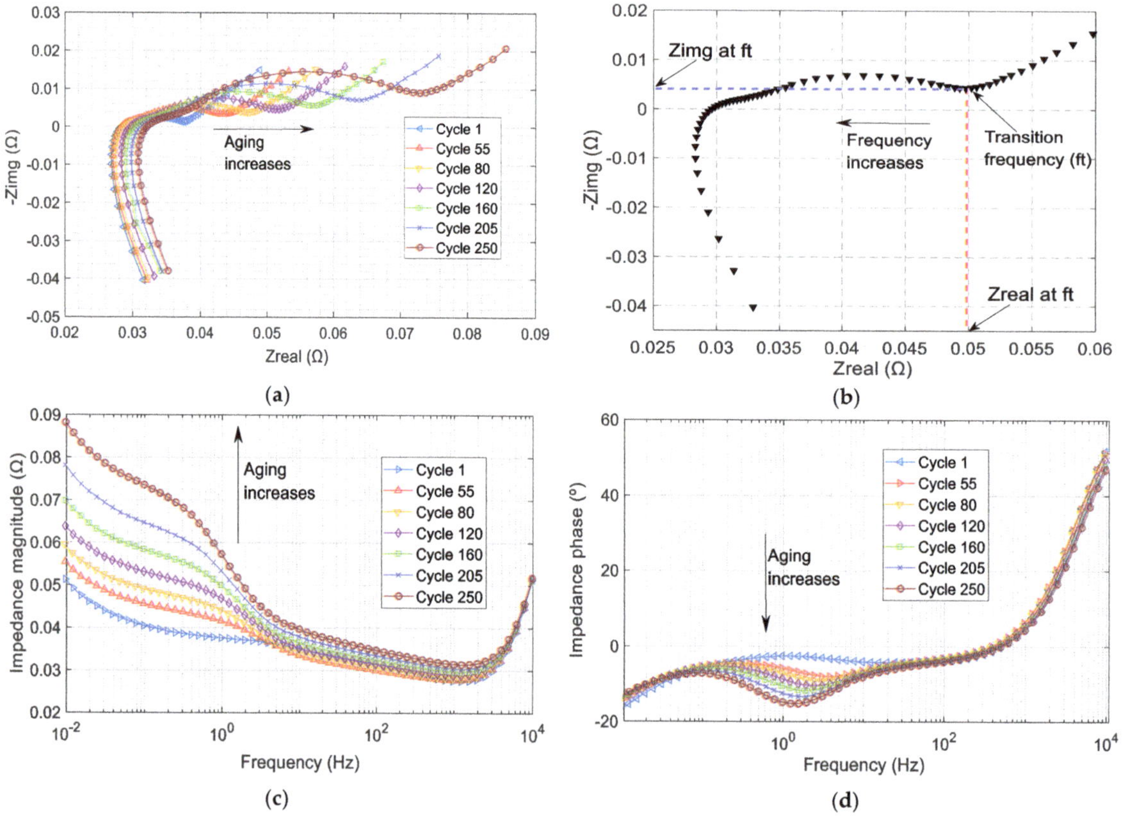

Figure 4. Impedance plots at different aging levels. (**a**) Nyquist plots. (**b**) Illustration of the impedance real and imaginary parts at the transition frequency. (**c**) Impedance magnitude curves. (**d**) Impedance phase curves.

(1) $SOH_{EST\#1}$: This SOH estimator utilizes two indicators obtained from the complex impedance Nyquist plot. The two indicators are the impedance real part Re $\{Z_{ft}\}$ and the impedance imaginary part Img $\{Z_{ft}\}$ at the transition frequency f_t. The transition frequency is the frequency at which the Nyquist plot moves from the diffusion region to the charge transfer region in the low-frequency range. Figure 4a shows sample Nyquist plots of BAT#1 at different aging cycles. As shown, as the battery ages (as the SOH value decreases), the Nyquist plot moves to the right (which means a larger Re $\{Z_{ft}\}$) and moves up (which means a larger Img $\{Z_{ft}\}$). A single Nyquist plot is shown in Figure 4b with the transition frequency f_t, Re $\{Z_{ft}\}$, and Img $\{Z_{ft}\}$ marked on the figure.

(2) $SOH_{EST\#2}$ and $SOH_{EST\#3}$: These estimators utilize certain impedance magnitude and phase values (respectively) within key frequency ranges in which the impedance magnitude and phase show strong correlations with the SOH value. From Figure 4c, it can be noticed that the impedance magnitude curves show a strong correlation with SOH values, especially in the low-frequency range. Therefore, impedance magnitude values at certain frequency points can be utilized for $SOH_{EST\#2}$. Similarly, from Figure 4d, it can be noticed that the impedance phase curves at different aging levels show a strong correlation with the SOH values in the low-frequency range. This means that impedance phase values at certain frequency points can be utilized for $SOH_{EST\#3}$.

4. Correlation Analysis

To measure the degree to which the SOH value changes with different SOH indicators and to identify the indicators that exhibit a strong correlation with SOH values, a correlation analysis is performed. The Spearman correlation coefficient is selected to perform the correlation analysis. The Spearman correlation coefficient (ρ) evaluates the monotonic relationship between two variables and is expressed in (2) [50]:

$$\rho = 1 - \frac{6\sum_{i=1}^{m} d_i^2}{m(m^2-1)} \tag{2}$$

where d is the rank difference of each observation and m is the number of observations. To put this into the context of the battery aging data in this paper, the matrix Z_{ft} in (3) represents the transition frequency's impedance real and imaginary values at m different SOH values. The matrix Z_{ft} has m rows (the number of SOH values obtained from battery aging) and two columns (one for Re $\{Z_{ft}\}$ values and another for Img $\{Z_{ft}\}$ values). The matrix $|Z|$ in (4) represents the impedance magnitude values over a frequency range of 0.01 Hz to 10 kHz at m different SOH values. The matrix $|Z|$ has m rows (the number of SOH values obtained from battery aging) and 61 columns (the number of impedance spectrum frequency points from 0.01 Hz to 10 kHz with 10 points per decade). Similarly, the matrix $\angle Z$ in (5) represents the impedance phase values over a frequency range of 0.01 Hz to 10 kHz at m different SOH values. The goal is to study the relationship between the values in each column of the matrices (3)–(5) and the SOH values. For example, to study the relationship between Re $\{Z_{ft}\}$ and the SOH value. This means studying the relationship between the first column in the Z_{ft} matrix expressed in (3) (Re $\{Z_{ft1}\}$, Re $\{Z_{ft2}\}$, Re $\{Z_{ft3}\}$ … Re $\{Z_{ftm}\}$) and the SOH values vector (SOH_1, SOH_2, SOH_3 … SOH_m) that is expressed in (6), which includes the SOH values obtained during aging after each cycle. The same procedure is applied for the matrices $|Z|$ and $\angle Z$. For example, studying the relationship between the impedance magnitudes at 0.01 Hz and the SOH value. This means studying the relationship between the first column in the $|Z|$ matrix ($|Z|_{1,0.01\,Hz}$, $|Z|_{2,0.01\,Hz}$, $|Z|_{3,0.01\,Hz}$ … $|Z|_{m,0.01\,Hz}$) and the SOH vector in (SOH_1, SOH_2, SOH_3 … SOH_m):

$$Z_{ft} = \begin{bmatrix} Re\{Z_{ft1}\} & Img\{Z_{ft1}\} \\ Re\{Z_{ft2}\} & Img\{Z_{ft2}\} \\ \vdots & \vdots \\ Re\{Z_{ftm}\} & Img\{Z_{ftm}\} \end{bmatrix} \tag{3}$$

$$|Z| = \begin{bmatrix} |Z|_{1,0.01Hz} & \cdots & |Z|_{1,10kHz} \\ |Z|_{2,0.01Hz} & \cdots & |Z|_{2,10kHz} \\ \vdots & \vdots & \vdots \\ |Z|_{m,0.01Hz} & \cdots & |Z|_{m,\,10kHz} \end{bmatrix} \tag{4}$$

$$\angle Z = \begin{bmatrix} \angle Z_{1,0.01Hz} & \cdots & \angle Z_{1,10kHz} \\ \angle Z_{2,0.01Hz} & \cdots & \angle Z_{2,10kHz} \\ \vdots & \vdots & \vdots \\ \angle Z_{m,0.01Hz} & \cdots & \angle Z_{m,10kHz} \end{bmatrix} \tag{5}$$

$$SOH = \begin{bmatrix} SOH_1 \\ SOH_2 \\ \vdots \\ SOH_m \end{bmatrix} \tag{6}$$

To find the rank difference d of each observation, the values in each column in Z_{ft}, $|Z|$, and $\angle Z$ are sorted in ascending order. Also, the SOH values in the SOH vector are sorted in ascending order. Then, the rank difference is calculated by finding the difference in each value's rank and the rank of its corresponding SOH value. For example, the values of Re $\{Z_{ft}\}$ [first column in (3)] are sorted in ascending order. Then, the rank difference d is calculated by finding the difference between the rank of each Re $\{Z_{ft}\}$ value and the rank of its corresponding SOH value. The same procedure is applied to the second column in (3), which includes the values for Img $\{Z_{ft}\}$. Also, the same procedure is applied to every column in the impedance magnitude matrix in (4) and to every column in the impedance phase matrix in (5). In other words, the correlation coefficient is calculated between the SOH vector in (6) and every single column in (3)–(5). The Spearman correlation coefficient can take values between -1 (a perfect negative monotonic association) to +1 (a perfect positive monotonic association). The Spearman correlation coefficient is calculated for the Re $\{Z_{ft}\}$ and Img $\{Z_{ft}\}$ values at the transition frequency; the impedance magnitude value at every frequency point; and the impedance phase value at every frequency point. The Spearman correlation coefficients for Re $\{Z_{ft}\}$ and Img $\{Z_{ft}\}$ are summarized in Table 2. Both the Re $\{Z_{ft}\}$ and Img $\{Z_{ft}\}$ impedance values at the transition frequency show strong correlations with the SOH and thus can be used as input features for $SOH_{EST\#1}$. Based on the aging data used in this paper, the transition frequency ranges from 1.585 Hz to 0.079 Hz at an SOH of ~96% and ~60%, respectively.

Table 2. Spearman correlation coefficients for Re $\{Z_{ft}\}$ and Img $\{Z_{ft}\}$.

	Re $\{Z_{ft}\}$	**Img $\{Z_{ft}\}$**
Battery#1	−0.9996	−0.9997
Battery#2	−0.9995	−0.9996
Battery#3	−0.9993	−0.9993

The Spearman correlation coefficients for the impedance magnitude values and impedance phase values are shown in Figure 5. From Figure 5a, it can be noticed that the impedance magnitude values at the lower frequency range of 0.01 Hz to 2 Hz show strong correlations with the SOH values, as shown in the zoomed-in view in Figure 5a. This means that impedance magnitude values within this frequency range are suitable SOH features for $SOH_{EST\#2}$ (which utilizes impedance magnitude values at specific frequency points/ranges). From Figure 5b, the impedance phase values show strong correlations with the SOH values within two frequency ranges, one around 1 Hz and another between 10 Hz and 20 Hz, as shown in the zoomed-in view in Figure 5b. This means that impedance phase values within these two frequency ranges can be used as input features for $SOH_{EST\#3}$ (which utilizes impedance phase values at specific frequency points/ranges).

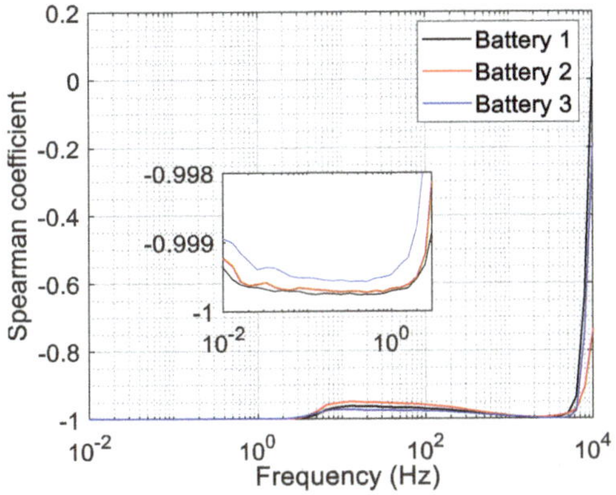

(a)

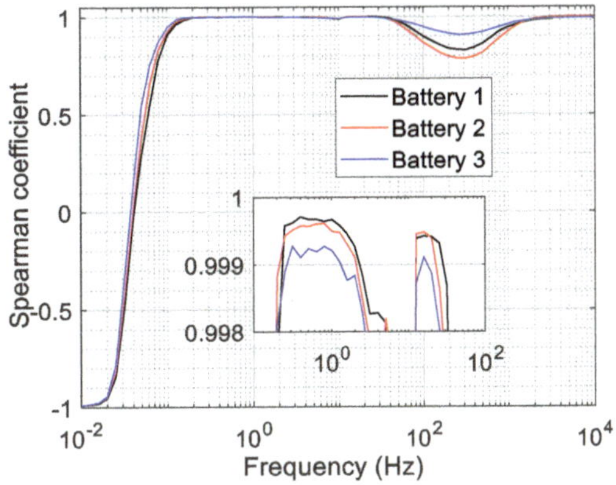

(b)

Figure 5. Spearman correlation coefficients between battery SOH values and the (**a**) impedance magnitude values, (**b**) impedance phase values.

5. ANN-Based SOH Estimation

The three SOH estimators explained in Section 3 are implemented using an artificial neural network (ANN). For $SOH_{EST\#1}$, there are two input features that are Re $\{Z_{ft}\}$ and Img $\{Z_{ft}\}$ values at the transition frequency obtained from the impedance Nyquist plot. For $SOH_{EST\#2}$, although many low-frequency points can be used as input features for the estimator based on the correlation analysis in the previous section, only the frequency points at which the impedance magnitude has the three largest absolute correlation coefficients are chosen as input features. This is to simplify the estimation model and to reduce its computational burden. Following the same intuition for $SOH_{EST\#3}$, the impedance phase values of the three strongest correlations with SOH value are utilized to train and test $SOH_{EST\#3}$.

Because the input features have different scales, the input data should be standardized, which can be applied as in (7):

$$x' = \frac{x - \mu}{\sigma} \tag{7}$$

where x' represents the standardized data point, x is the original data point, μ is the input data mean, and σ is the input data standard deviation.

After normalizing the input features, the training dataset is split into k-fold cross-validation sets. The k-fold cross-validation [51] is utilized to guide the SOH estimation models' training and can be utilized as follows: for each fold of the training dataset, the model is trained on a part of the fold data points and tested/validated on the remainder of the fold data points. This process is repeated k times (i.e., k folds). For every fold, the validation error (the mean square error or "MSE") is recorded, and if the average validation error of all training sessions is less than a pre-specified threshold, then the training process is considered complete, and the model becomes ready to be tested on the testing dataset that the model has never seen before. If the average validation error is not below the pre-specified threshold, the neural network hyperparameters need to be changed (e.g., changing the number of neurons in the hidden layer) until the average validation error becomes smaller than the pre-specified threshold. In this paper, the threshold for the average validation error is set to be 125 Coulombs (C), which is equivalent to ~0.035 Ah (125 C × 1 h/3600 s). This is equivalent to an SOH estimation error of 1.0% (0.035 Ah/3.5 Ah = 1.0%). The training dataset for each estimator is split into 10 folds (k = 10). Based on the k-fold cross-validation analysis, a three-layer ANN is utilized (input layer, hidden layer, and output layer) for the three SOH estimators. $SOH_{EST\#1}$ utilizes an input layer of two neurons ($SOH_{EST\#1}$ utilizes two features that are Re $\{Z_{ft}\}$ and Img $\{Z_{ft}\}$) and a hidden layer of eight neurons. $SOH_{EST\#2}$ utilizes an input layer of three neurons ($SOH_{EST\#2}$ utilizes three impedance magnitude values) and a hidden layer of twelve neurons. $SOH_{EST\#3}$ utilizes an input layer of three neurons ($SOH_{EST\#3}$ utilizes three impedance phase values) and a hidden layer of nine neurons. For all SOH estimators in this paper, one neuron is utilized in the output layer to give the estimated SOH value.

The overall framework for the SOH estimators $SOH_{EST\#1}$, $SOH_{EST\#2}$, and $SOH_{EST\#3}$ is shown in Figure 6. Before feeding the SOH features into the neural network, the input features go through a preprocessing stage. Data preprocessing includes data cleansing (removing any outliers that might exist), feature extraction using correlation analysis, and standardization [expressed in (7)].

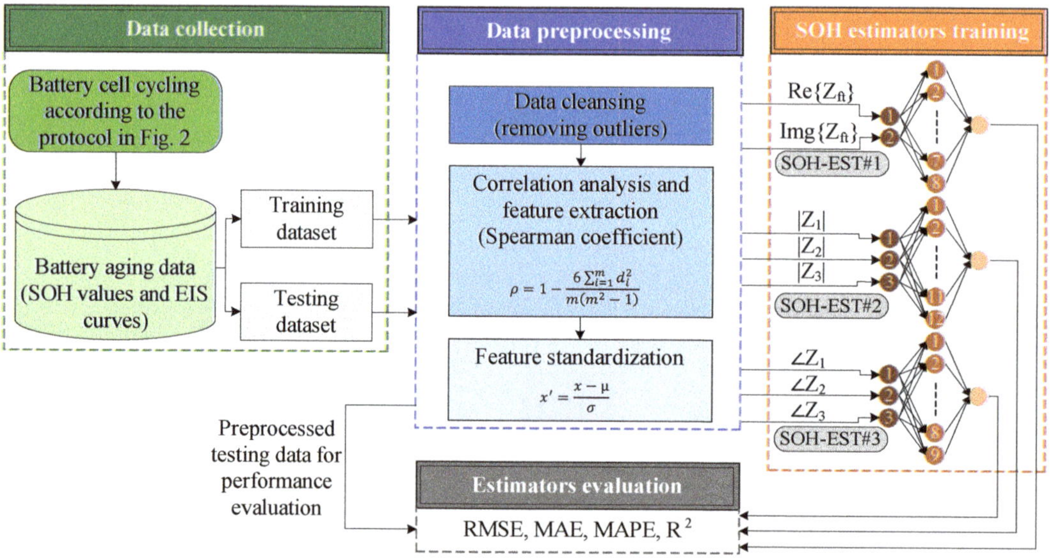

Figure 6. Overall SOH estimation framework.

The SOH estimators are trained, validated, and evaluated as follows: aging data (EIS curves and SOH values) for two batteries is used for SOH estimator training, and the estimator performance is evaluated using the aging data of the remaining battery (the SOH estimator performance is evaluated on a dataset that was not included in training). For example, aging data for BAT#1 and BAT#2 are used to train and validate the SOH estimator, where the SOH estimator's performance is evaluated using the aging data of BAT#3. This way, there are multiple variations in the SOH estimator training and testing datasets. These variations are summarized in Table 3. For example, $SOH_{EST\#1\_12\_3}$ represents the case in which $SOH_{EST\#1}$ is trained using the aging data of BAT#1 and BAT#2 (i.e., "12" in the subscript of $SOH_{EST\#1\_12\_3}$) where the performance of the SOH estimator is evaluated on the aging data of BAT#3 (i.e., "3" in the subscript of $SOH_{EST\#1\_12\_3}$).

Table 3. SOH estimator training, validation, and evaluation datasets summary.

SOH Estimator	SOH Estimation Model Designator	Training and Validation Data (Battery Numbers)	Performance Evaluation Data (Battery Number)	Frequency Points of Largest Correlation Coefficients
$SOH_{EST\#1}$	$SOH_{EST\#1\_12\_3}$	1, 2	3	Transition frequency ranges between 1.585 Hz and 0.079 Hz
	$SOH_{EST\#1\_13\_2}$	1, 3	2	
	$SOH_{EST\#1\_23\_1}$	2, 3	1	
$SOH_{EST\#2}$	$SOH_{EST\#2\_12\_3}$	1, 2	3	0.079 Hz, 0.158 Hz, 0.2 Hz
	$SOH_{EST\#2\_13\_2}$	1, 3	2	0.316 Hz, 0.398 Hz, 0.501 Hz
	$SOH_{EST\#2\_23\_1}$	2, 3	1	1.259 Hz, 1.585 Hz, 1.995 Hz
$SOH_{EST\#3}$	$SOH_{EST\#3\_12\_3}$	1, 2	3	12.59 Hz, 15.85 Hz, 19.95 Hz
	$SOH_{EST\#3\_13\_2}$	1, 3	2	12.59 Hz, 15.85 Hz, 19.95 Hz
	$SOH_{EST\#3\_23\_1}$	2, 3	1	12.59 Hz, 15.85 Hz, 19.95 Hz

It should be mentioned that for $SOH_{EST\#2}$, the frequency points of the correlation between the impedance magnitude and the SOH are not necessarily the same among all batteries. For example, for $SOH_{EST\#2\_12\_3}$, the correlation analysis is performed on the data of BAT#1 and BAT#2 to identify the three frequency points at which the correlations between the impedance magnitude and the SOH are the largest. This analysis resulted in identifying 0.079 Hz, 0.158 Hz, and 0.2 Hz as the three frequency points of interest. However, this is different from those of $SOH_{EST\#2\_23\_1}$ and $SOH_{EST\#1\_13\_2}$. The frequency points of interest for each estimator are summarized in Table 3 based on the correlation analysis of each training dataset. It should be mentioned that some of the frequency points at which the real and imaginary parts at the transition frequency are utilized for $SOH_{EST\#1}$ might be the same frequency points at which the $SOH_{EST\#2}$ features are extracted based on the correlation analysis, as shown in Table 3. However, this paper presents each of the SOH estimators separately and compares them.

The mean square error (MSE) is used as a loss function in the ANN training as expressed in (8), where $y(i)$ is the value of the actual SOH in the training dataset and $y'(i)$ is the value of the estimated SOH for the ith training sample:

$$MSE = \frac{1}{n}\sum_{i=1}^{n}|y(i) - y\prime(i)| \tag{8}$$

To evaluate the estimation accuracy of the SOH estimators, the root mean square error (RMSE), the mean absolute error (MAE), the mean absolute percentage error (MAPE), and the coefficient of determination (R^2) are utilized. These four indices are expressed in (9)–(12), where n is the number of outputs (the SOH values), A(i) is the actual output (the true SOH value), and P(i) is the estimated output (the estimated SOH value):

$$RMSE = \sqrt{\frac{1}{n}\sum_{i=1}^{n}(A(i) - P(i))^2} \tag{9}$$

$$MAE = \frac{1}{n}\sum_{i=1}^{n}|A(i) - P(i)| \tag{10}$$

$$MAPE = \frac{100}{n}\sum_{i=1}^{n}\frac{|A(i) - P(i)|}{A(i)} \tag{11}$$

$$R^2 = 1 - \frac{SS_{res}}{SS_{tot}} = 1 - \frac{\sum_{i=1}^{n}(A(i) - P(i))^2}{\sum_{i=1}^{n}\left(A(i) - \overline{A(i)}\right)^2} \tag{12}$$

In (12), the term SS_{res} represents the total sum of squared residuals between the actual SOH values and the predicted SOH values. The term SS_{tot} represents the total sum of squares, i.e., the sum of squared differences between the actual SOH values and their mean value $\overline{A(i)}$.

Following the ANN model training details in this section, the SOH estimators' performance indicators are summarized in Table 4. Also, in Figure 7, the true SOH values versus the estimated SOH values are plotted, where the black dashed line represents the ideal estimation case (i.e., as if the estimated SOH value is equal to the actual SOH value).

Table 4. SOH estimators' performance results.

SOH Estimator	SOH Estimation Model Designator	RMSE (%)	MAE (%)	MAPE (%)	R^2
$SOH_{EST\#1}$	$SOH_{EST\#1\_12\_3}$	1.76	1.30	1.94	0.978
	$SOH_{EST\#1\_13\_2}$	1.24	1.08	1.39	0.986
	$SOH_{EST\#1\_23\_1}$	0.67	0.54	0.68	0.995
	Average	1.22	0.97	1.34	0.986
$SOH_{EST\#2}$	$SOH_{EST\#2\_12\_3}$	1.18	0.98	1.34	0.990
	$SOH_{EST\#2\_13\_2}$	1.40	1.29	1.66	0.986
	$SOH_{EST\#2\_23\_1}$	1.60	1.47	1.94	0.973
	Average	1.39	1.25	1.55	0.983
$SOH_{EST\#3}$	$SOH_{EST\#3\_12\_3}$	0.98	0.81	0.76	0.993
	$SOH_{EST\#3\_13\_2}$	0.74	0.60	0.90	0.996
	$SOH_{EST\#3\_23\_1}$	0.82	0.69	0.98	0.993
	Average	0.85	0.70	0.88	0.994

From Table 4 and Figure 7, it is clear that all SOH estimators are capable of accurately estimating the SOH values based on their input features. The estimated SOH values match well with the actual SOH values. Table 4 also includes the average values of the RMSE, MAE, MAPE, and R^2 to make a comparison between the three estimators easier. The average values for all performance indicators are used to compare the three SOH estimators against each other, as shown in Figure 8. It can be noticed that $SOH_{EST\#3}$ (which utilizes the impedance phase values at specific frequency points) shows the best performance among the three estimators; it has the lowest RMSE, MAE, and MAPE, in addition to the highest R^2. The second-best estimator is $SOH_{EST\#1}$ (which utilizes the real and imaginary parts of the impedance at the transition frequency).

While multiple SOH estimators are presented in this paper, it is not necessary to use all of them. One estimator should be enough to predict the SOH value. Also, based on the analysis presented in this paper, it is not necessary to measure the entire EIS spectrum for any of the estimators. For $SOH_{EST\#1}$, it would be only required to obtain the EIS data over a specific frequency range (between 1.585 Hz and 0.079 Hz). For $SOH_{EST\#2}$ and $SOH_{EST\#3}$, it would be only required to obtain the impedance magnitude and phase values at the frequency points that show the strongest correlations with SOH values. For instance, obtaining the impedance phase at 12.59 Hz, 15.85 Hz, and 19.95 Hz for $SOH_{EST\#3}$. Another issue to be taken into consideration is the time taken to perform the EIS measurement. For example, obtaining the impedance magnitude/phase at a very low frequency (e.g., 0.02 Hz) takes more time than obtaining the EIS measurement at 20 Hz. This gives $SOH_{EST\#3}$ an advantage over $SOH_{EST\#2}$ and $SOH_{EST\#1}$ in terms of speed since $SOH_{EST\#3}$ utilizes impedance phase values at 12.59 Hz, 15.85 Hz, and 19.95 Hz (which takes ~6 s to measure), which are at a higher frequency range (i.e., a faster EIS measurement) compared with $SOH_{EST\#2}$ (between 0.079 Hz and 0.2 Hz, which takes ~48 s to measure) and $SOH_{EST\#1}$ (which requires measuring the EIS between 1.585 Hz and 0.079 Hz and that takes ~2 min). For all the estimators, the time taken to perform the EIS measurement is still significantly lower (a maximum of ~2 min) than the time needed to measure the entire EIS curve (~17 min), which is equivalent to an ~88% reduction in measurement time.

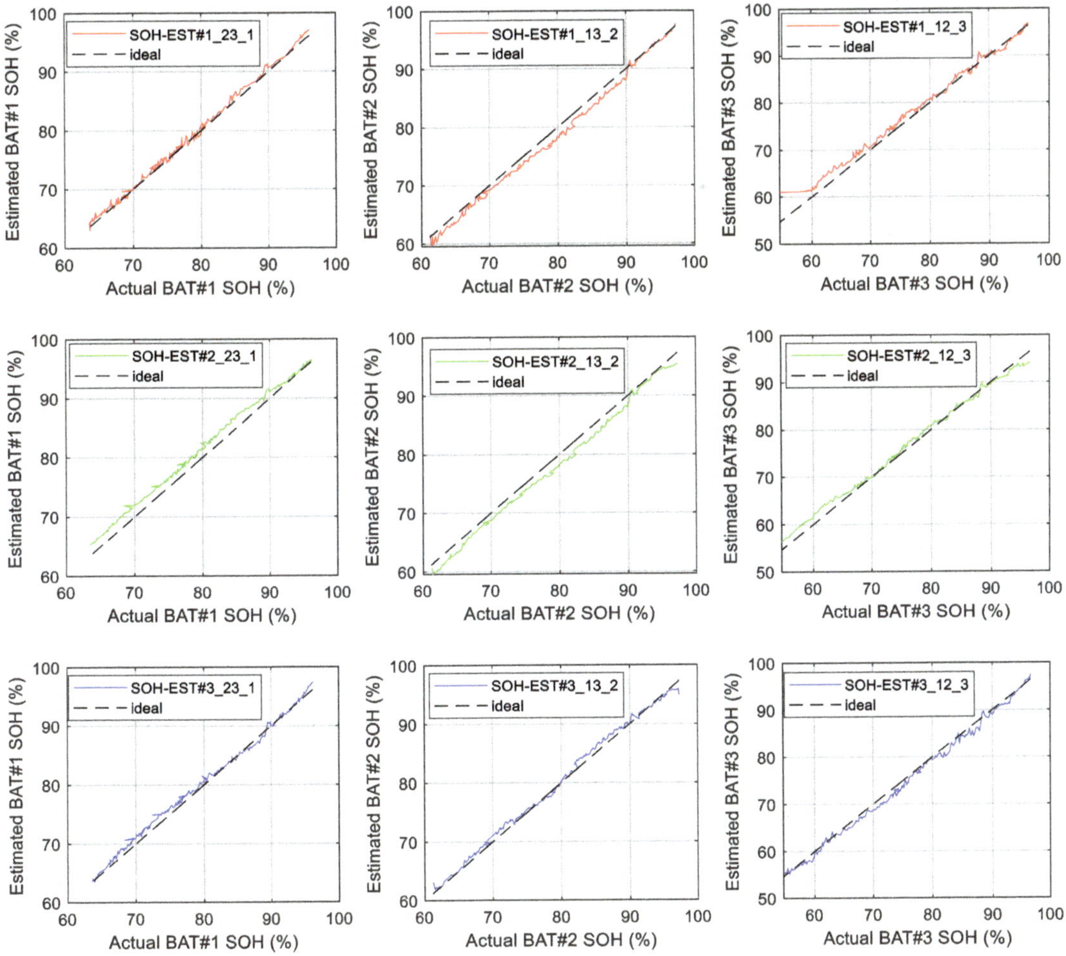

Figure 7. Actual vs estimated SOH for $SOH_{EST\#1}$ (**top**), $SOH_{EST\#2}$ (**middle**), and $SOH_{EST\#3}$ (**bottom**).

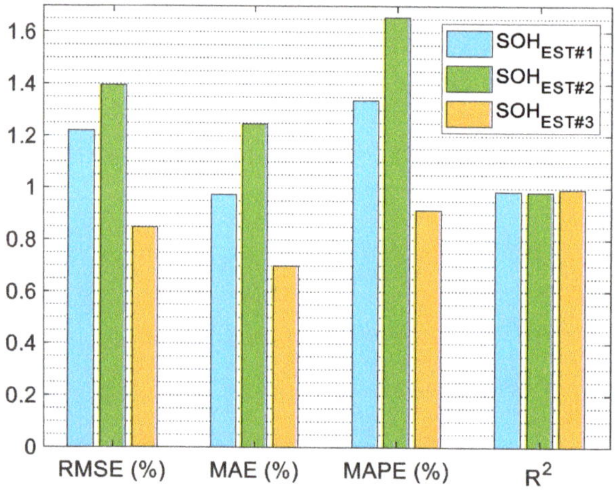

Figure 8. Performance comparison between different SOH estimators.

6. Conclusions

This paper presented three EIS-based SOH estimators for lithium-ion batteries by utilizing different features obtained from the impedance Nyquist plot, impedance magnitude curve, and impedance phase curve. This article (a) presented new health features obtained from the Nyquist plot for SOH estimation, (b) presented a correlation-analysis-based framework to obtain impedance magnitude and impedance phase values to be utilized for SOH estimation, and (c) presented and evaluated ANN-based SOH estimators. The performance evaluation of the three developed SOH estimators shows that all estimators are capable of accurately predicting the SOH values with minor/negligible differences in their accuracies.

Future work includes, but is not limited to, testing the presented SOH estimators on larger battery aging datasets of high degradation levels, in addition to comparing the presented estimation methods with other methods in the literature, such as Kalman filter and particle filter-based methods. Also, different types of batteries can be considered (different brands and different capacities) to enhance the generalizability of SOH estimation models/techniques.

Author Contributions: Conceptualization, M.K.A.-S. and J.A.A.Q.; methodology, M.K.A.-S. and J.A.A.Q.; validation, M.K.A.-S.; formal analysis, M.K.A.-S.; investigation, M.K.A.-S. and J.A.A.Q.; resources, J.A.A.Q.; data curation, M.K.A.-S.; writing—original draft preparation, M.K.A.-S.; writing—review and editing, J.A.A.Q.; visualization, M.K.A.-S. and J.A.A.Q.; supervision, J.A.A.Q.; project administration, J.A.A.Q.; funding acquisition, J.A.A.Q. All authors have read and agreed to the published version of the manuscript.

Funding: This material is based on work supported, in part, by the National Science Foundation (NSF, United States) under grant no. 2213918. Any opinions, findings, and conclusions or recommendations expressed in this material are those of the author(s) and do not necessarily reflect the views of the National Science Foundation.

Data Availability Statement: The original contributions presented in the study are included in the article; further inquiries can be directed to the author.

Conflicts of Interest: The authors declare no conflicts of interest.

References

1. Manzetti, S.; Mariasiu, F. Electric vehicle battery technologies: From present state to future systems. *Renew. Sustain. Energy Rev.* **2015**, *51*, 1004–1012. [CrossRef]
2. Deng, Z.; Hu, X.; Li, P.; Lin, X.; Bian, X. Data-Driven Battery State of Health Estimation Based on Random Partial Charging Data. *IEEE Trans. Power Electron.* **2022**, *37*, 5021–5031. [CrossRef]
3. Xia, Z.; Qahouq, J.A.A. Lithium-ion battery ageing behavior pattern characterization and state-of-health estimation using data-driven method. *IEEE Access* **2021**, *9*, 98287–98304.
4. Lee, J.; Kwon, D.; Pecht, M.G. Reduction of li-ion battery qualification time based on prognostics and health management. *IEEE Trans. Ind. Electron.* **2018**, *66*, 7310–7315.
5. Soyoye, B.D.; Bhattacharya, I.; Anthony Dhason, M.V.; Banik, T. State of Charge and State of Health Estimation in Electric Vehicles: Challenges, Approaches and Future Directions. *Batteries* **2025**, *11*, 32. [CrossRef]
6. Nuroldayeva, G.; Serik, Y.; Adair, D.; Uzakbaiuly, B.; Bakenov, Z. State of Health Estimation Methods for Lithium-Ion Batteries. *Int. J. Energy Res.* **2023**, *2023*, 4297545.
7. Akram, M.N.; Abdul-Kader, W. Repurposing Second-Life EV Batteries to Advance Sustainable Development: A Comprehensive Review. *Batteries* **2024**, *10*, 452. [CrossRef]
8. Zhao, Y.; Qahouq, J.A.A. Second-Use Battery System for EV Charging Stations. In Proceedings of the 2023 IEEE Energy Conversion Congress and Exposition (ECCE), Nashville, TN, USA, 29 October–2 November 2023; pp. 2122–2126.
9. Vignesh, S.; Che, H.S.; Selvaraj, J.; Tey, K.S.; Lee, J.W.; Shareef, H.; Errouissi, R. State of Health (SoH) estimation methods for second life lithium-ion battery—Review and challenges. *Appl. Energy* **2024**, *369*, 123542.
10. Vermeer, W.; Mouli, G.R.C.; Bauer, P. A comprehensive review on the characteristics and modeling of lithium-ion battery aging. *IEEE Trans. Transp. Electrif.* **2021**, *8*, 2205–2232.
11. Han, X.; Lu, L.; Zheng, Y.; Feng, X.; Li, Z.; Li, J.; Ouyang, M. A review on the key issues of the lithium ion battery degradation among the whole life cycle. *ETransportation* **2019**, *1*, 100005.
12. Kumar, R.R.; Bharatiraja, C.; Udhayakumar, K.; Devakirubakaran, S.; Sekar, K.S.; Mihet-Popa, L. Advances in Batteries, Battery Modeling, Battery Management System, Battery Thermal Management, SOC, SOH, and Charge/Discharge Characteristics in EV Applications. *IEEE Access* **2023**, *11*, 105761–105809. [CrossRef]
13. Ahwiadi, M.; Wang, W. Battery Health Monitoring and Remaining Useful Life Prediction Techniques: A Review of Technologies. *Batteries* **2025**, *11*, 31. [CrossRef]
14. Yang, B.; Qian, Y.; Li, Q.; Chen, Q.; Wu, J.; Luo, E.; Xie, R.; Zheng, R.; Yan, Y.; Su, S. Critical summary and perspectives on state-of-health of lithium-ion battery. *Renew. Sustain. Energy Rev.* **2024**, *190*, 114077.
15. Demirci, O.; Taskin, S.; Schaltz, E.; Demirci, B.A. Review of battery state estimation methods for electric vehicles-Part II: SOH estimation. *J. Energy Storage* **2024**, *96*, 112703.
16. Yao, L.; Xu, S.; Tang, A.; Zhou, F.; Hou, J.; Xiao, Y.; Fu, Z. A review of lithium-ion battery state of health estimation and prediction methods. *World Electr. Veh. J.* **2021**, *12*, 113. [CrossRef]
17. Yang, S.; Zhang, C.; Jiang, J.; Zhang, W.; Zhang, L.; Wang, Y. Review on state-of-health of lithium-ion batteries: Characterizations, estimations and applications. *J. Clean. Prod.* **2021**, *314*, 128015.
18. Guha, A.; Patra, A. State of health estimation of lithium-ion batteries using capacity fade and internal resistance growth models. *IEEE Trans. Transp. Electrif.* **2017**, *4*, 135–146.
19. Zhang, C.; Liu, J.; Sharkh, S. Identification of dynamic model parameters for lithium-ion batteries used in hybrid electric vehicles. In Proceedings of the International Symposium on Electric Vehicles (ISEV), Beijing, China, 1 September 2009.
20. Bao, Y.; Dong, W.; Wang, D. Online Internal Resistance Measurement Application in Lithium Ion Battery Capacity and State of Charge Estimation. *Energies* **2018**, *11*, 1073. [CrossRef]
21. Fu, Y.; Xu, J.; Shi, M.; Mei, X. A fast impedance calculation-based battery state-of-health estimation method. *IEEE Trans. Ind. Electron.* **2021**, *69*, 7019–7028.
22. Wang, D.; Bao, Y.; Shi, J. Online lithium-ion battery internal resistance measurement application in state-of-charge estimation using the extended Kalman filter. *Energies* **2017**, *10*, 1284. [CrossRef]
23. Movassagh, K.; Raihan, A.; Balasingam, B.; Pattipati, K. A Critical Look at Coulomb Counting Approach for State of Charge Estimation in Batteries. *Energies* **2021**, *14*, 4074. [CrossRef]
24. Schaltz, E.; Stroe, D.-I.; Nørregaard, K.; Ingvardsen, L.S.; Christensen, A. Incremental capacity analysis applied on electric vehicles for battery state-of-health estimation. *IEEE Trans. Ind. Appl.* **2021**, *57*, 1810–1817. [CrossRef]
25. Xia, F.; Wang, K.; Chen, J. State of health and remaining useful life prediction of lithium-ion batteries based on a disturbance-free incremental capacity and differential voltage analysis method. *J. Energy Storage* **2023**, *64*, 107161. [CrossRef]

26. Huang, Z.; Best, M.; Knowles, J.; Fly, A. Adaptive Piecewise Equivalent Circuit Model With SOC/SOH Estimation Based on Extended Kalman Filter. *IEEE Trans. Energy Convers.* **2023**, *38*, 959–970. [CrossRef]
27. Andre, D.; Appel, C.; Soczka-Guth, T.; Sauer, D.U. Advanced mathematical methods of SOC and SOH estimation for lithium-ion batteries. *J. Power Sources* **2013**, *224*, 20–27. [CrossRef]
28. Fang, F.; Ma, C.; Ji, Y. A Method for State of Charge and State of Health Estimation of Lithium Batteries Based on an Adaptive Weighting Unscented Kalman Filter. *Energies* **2024**, *17*, 2145. [CrossRef]
29. Liu, D.; Yin, X.; Song, Y.; Liu, W.; Peng, Y. An On-Line State of Health Estimation of Lithium-Ion Battery Using Unscented Particle Filter. *IEEE Access* **2018**, *6*, 40990–41001. [CrossRef]
30. Wu, T.; Liu, S.; Wang, Z.; Huang, Y. SOC and SOH joint estimation of lithium-ion battery based on improved particle filter algorithm. *J. Electr. Eng. Technol.* **2022**, *17*, 307–317. [CrossRef]
31. Bi, J.; Zhang, T.; Yu, H.; Kang, Y. State-of-health estimation of lithium-ion battery packs in electric vehicles based on genetic resampling particle filter. *Appl. Energy* **2016**, *182*, 558–568. [CrossRef]
32. Zhang, Z.; Zhang, R.; Liu, X.; Zhang, C.; Sun, G.; Zhou, Y.; Yang, Z.; Liu, X.; Chen, S.; Dong, X.; et al. Advanced State-of-Health Estimation for Lithium-Ion Batteries Using Multi-Feature Fusion and KAN-LSTM Hybrid Model. *Batteries* **2024**, *10*, 433. [CrossRef]
33. Al-Smadi, M.K.; Qahouq, J.A.A. SOH Estimation Algorithm and Hardware Platform for Lithium-ion Batteries. In Proceedings of the 2024 IEEE Vehicle Power and Propulsion Conference (VPPC), Washington DC, USA, 7–10 October 2024; pp. 1–5.
34. Abu Qahouq, J. An Electrochemical Impedance Spectrum-Based State of Health Differential Indicator with Reduced Sensitivity to Measurement Errors for Lithium–Ion Batteries. *Batteries* **2024**, *10*, 368. [CrossRef]
35. Zhang, Y.; Tang, Q.; Zhang, Y.; Wang, J.; Stimming, U.; Lee, A.A. Identifying degradation patterns of lithium ion batteries from impedance spectroscopy using machine learning. *Nat. Commun.* **2020**, *11*, 1706. [PubMed]
36. Xia, Z.; Qahouq, J.A.A. State of Health Estimation of Lithium-Ion Batteries Using Neuron Network and 1kHz Impedance Data. In Proceedings of the 2020 IEEE Energy Conversion Congress and Exposition (ECCE), Detroit, MI, USA; 2020; pp. 1968–1972.
37. Love, C.T.; Dubarry, M.; Reshetenko, T.; Devie, A.; Spinner, N.; Swider-Lyons, K.E.; Rocheleau, R. Lithium-ion cell fault detection by single-point impedance diagnostic and degradation mechanism validation for series-wired batteries cycled at 0 C. *Energies* **2018**, *11*, 834. [CrossRef]
38. Eddahech, A.; Briat, O.; Woirgard, E.; Vinassa, J.M. Remaining useful life prediction of lithium batteries in calendar ageing for automotive applications. *Microelectron. Reliab.* **2012**, *52*, 2438–2442. [CrossRef]
39. Xia, Z.; Qahouq, J.A.A. State-of-charge balancing of lithium-ion batteries with state-of-health awareness capability. *IEEE Trans. Ind. Appl.* **2020**, *57*, 673–684. [CrossRef]
40. Xia, Z.; Qahouq, J.A.A. Adaptive and Fast State of Health Estimation Method for Lithium-ion Batteries Using Online Complex Impedance and Artificial Neural Network. In Proceedings of the 2019 IEEE Applied Power Electronics Conference and Exposition (APEC), Anaheim, CA, USA, 17–21 March 2019; pp. 3361–3365.
41. Li, C.; Yang, L.; Li, Q.; Zhang, Q.; Zhou, Z.; Meng, Y.; Zhao, X.; Wang, L.; Zhang, S.; Li, Y. SOH estimation method for lithium-ion batteries based on an improved equivalent circuit model via electrochemical impedance spectroscopy. *J. Energy Storage* **2024**, *86*, 111167.
42. Xiong, R.; Tian, J.; Mu, H.; Wang, C. A systematic model-based degradation behavior recognition and health monitoring method for lithium-ion batteries. *Appl. Energy* **2017**, *207*, 372–383.
43. Luo, F.; Huang, H.; Ni, L.; Li, T. Rapid prediction of the state of health of retired power batteries based on electrochemical impedance spectroscopy. *J. Energy Storage* **2021**, *41*, 102866.
44. Wang, X.; Wei, X.; Dai, H. Estimation of state of health of lithium-ion batteries based on charge transfer resistance considering different temperature and state of charge. *J. Energy Storage* **2019**, *21*, 618–631.
45. LGChem. *Product Specification: Rechargeable Lithium Ion Battery, Model: INR18650 MJ1 3500mAh*; LGChem: Seoul, Republic of Korea, 2024.
46. Admiral Instruments, "Squidstat Cycler—Base Model". Available online: https://www.admiralinstruments.com/_files/ugd/dc5bf5_c4286dd7c5be411c9ad415c452fe5c88.pdf (accessed on 13 March 2025).
47. Westerhoff, U.; Kroker, T.; Kurbach, K.; Kurrat, M. Electrochemical impedance spectroscopy based estimation of the state of charge of lithium-ion batteries. *J. Energy Storage* **2016**, *8*, 244–256.
48. Messing, M.; Shoa, T.; Habibi, S. Electrochemical impedance spectroscopy with practical rest-times for battery management applications. *IEEE Access* **2021**, *9*, 66989–66998. [CrossRef]
49. Li, Q.; Yi, D.; Dang, G.; Zhao, H.; Lu, T.; Wang, Q.; Lai, C.; Xie, J. Electrochemical Impedance Spectrum (EIS) Variation of Lithium-Ion Batteries Due to Resting Times in the Charging Processes. *World Electr. Veh. J.* **2023**, *14*, 321. [CrossRef]

50. Spearman, C. *The Proof and Measurement of Association Between Two Things*; Appleton-Century-Crofts: East Norwalk, CT, USA, 1961; pp. 45–58.
51. Berrar, D. Cross-Validation. In *Encyclopedia of Bioinformatics and Computational Biology*; Ranganathan, S., Gribskov, M., Nakai, K., Schönbach, C., Eds.; Academic Press: Oxford, UK, 2019; pp. 542–545.

batteries

MDPI

Article

Big Data Study of the Impact of Residential Usage and Inhomogeneities on the Diagnosability of PV-Connected Batteries

Fahim Yasir, Saeed Sepasi and Matthieu Dubarry *

Hawai'i Natural Energy Institute, University of Hawai'i at Mānoa, Honolulu, HI 96822, USA; fyasir@hawaii.edu (F.Y.); sepasi@hawaii.edu (S.S.)
* Correspondence: matthieu@hawaii.edu

Abstract: Grid-connected battery energy storage systems are usually used 24/7, which could prevent the utilization of typical diagnosis and prognosis techniques that require controlled conditions. While some new approaches have been proposed at the laboratory level, the impact of real-world conditions could still be problematic. This work investigates both the impact of additional residential usage on the cells while charging and of inhomogeneities on the diagnosability of batteries charged from photovoltaic systems. Using Big-Data synthetic datasets covering more than ten thousand possible degradations, we will show that these impacts can be accommodated to retain good diagnosability under auspicious conditions to reach average RMSEs around 2.75%.

Keywords: PV-connected batteries; Big Data; synthetic data; CNN; residential usage

1. Introduction

Harvesting energy from the Sun using solar panels requires significant storage to circumvent the natural daily and seasonal variations as well as the nocturnal cycle [1]. While large centralized PV farms associated with battery energy storage systems (BESS) will play a central role, they require a lot of land, which could be problematic for islands [2]. Distributed generation mitigates the land issue but adds several new problems, including the requirement for a significant number of smaller batteries to be connected to the grid, which drastically increases the risk of failure. Monitoring these batteries will require advanced state of health (SOH) estimation techniques [3,4] because the duty cycles will be sporadic without any opportunity for controlled charge or discharge steps. While data-driven methods [5–10] could offer a good solution to avoid downtime for diagnosis, their training could be challenging, as the degradation the cells will experience in the field is path-dependent [11–13] and largely unknown at the moment, although some data were recently made available [14]. Gathering enough data under varied enough conditions to replicate the wide range of possible uses for the cells and properly train algorithms will be both experimentally complex and expensive. For voltage-based diagnosis methods, a possible solution is to rely on synthetic datasets generated from different types of battery models [15]. Exemplary datasets covering the totality of the degradation spectrum are publicly available and have already been successfully applied in the training of machine learning algorithms [16–28]. However, their main limitation is that they are generated using controlled low-constant-current conditions, which are not representative of the usage of deployed batteries.

In recent years, we proposed a methodology to generate synthetic data outside of a constant current [29] and applied it to PV-connected batteries for nickel manganese cobalt

https://doi.org/10.3390/batteries11040154

oxide (NMC) [23,24] and lithium iron phosphate [25] chemistries to successfully diagnose the degradation using an algorithm trained on clear sky irradiance (CSI) using auspicious conditions. Since battery lifetime is supposed to be a decade or more, diagnosis does not have to be performed on a daily, or even weekly, basis, and our work showed that, in the example of a solar array located in Maui, USA, using only days with 50% or more clear sky will allow accurate diagnosis of the degradation, with an average root mean square error (RMSE) around 1.75%. However, this work also had its own limitations, as no additional usage of the generated power was considered throughout the battery charge and as the voltage response used for the simulation was the one of a single cell and not of a battery pack. While useful for proof of concept, this simplification is unrealistic as a home will be connected to a battery pack from which a significant amount of power will be used at different times of the day for normal usage.

This work's principal focus is to investigate the impact of these limitations and to propose mitigation strategies. Looking first into the impact of the additional usage, it is important to note that our approach involves using two distinct synthetic datasets for training and validation, as shown in Figure 1. The power estimated from a modeled CSI is used for training, whereas the power estimated from the observed irradiance (OI) is used for validation. For the validation cycles, the daily additional usage of the cells could simply be deducted from the power obtained from the daily observed irradiance. However, this cannot be done for the training dataset since, ultimately, the training will be completed before deployment and therefore without any knowledge of the actual daily usage the cells will encounter. To investigate possible mitigation strategies, three datasets were generated and used to train a machine learning algorithm that was tested on synthetic data consisting of the OI minus simulated residential load for any given day within one year (Figure 1). For training, the first dataset considered the CSI-generated power, similar to what we proposed previously [23–25] without any accommodation for the additional usage, the second dataset considered the CSI-generated power minus the average of the yearly residential loads (|Load|), and the third dataset consisted of the CSI-generated power minus the average of the yearly average residential load (||Load||), and thus a constant value throughout the day. In addition, and to further address the impact of the accuracy of the residential usage measurement, additional datasets were generated within plus or minus 20% around ||Load||. This was tested because the residual loads will be customer-dependent and thus hard to predict with accuracy without data monitoring.

For the second limitation, some studies have already showcased that the imbalance and cell-to-cell variations within a battery pack mostly increase the downward slope of the voltage vs. capacity curves, which correspond to a broadening of the derivative peaks [30–43]. Investigating this issue further, recent modeling work on simulations of nine different types of inhomogeneities [44], including loss of active material (LAM), resistance, state of charge (SOC), and kinetics, showcased that most should not affect much the voltage response of the pack. Considering the inhomogeneities that do affect the voltage response, the SOC ones should not be present in battery packs with proper balancing. This leaves only resistance inhomogeneities that were projected to broaden the derivate of the voltage response. This is consistent with literature observations, and it can be handled in our model by increasing kinetic limitations [29,45,46]. In this work, simulations were undertaken with three levels of kinetic limitations for training and five for validation in order to assess the impact of different levels of inhomogeneity on the accuracy of the diagnosis.

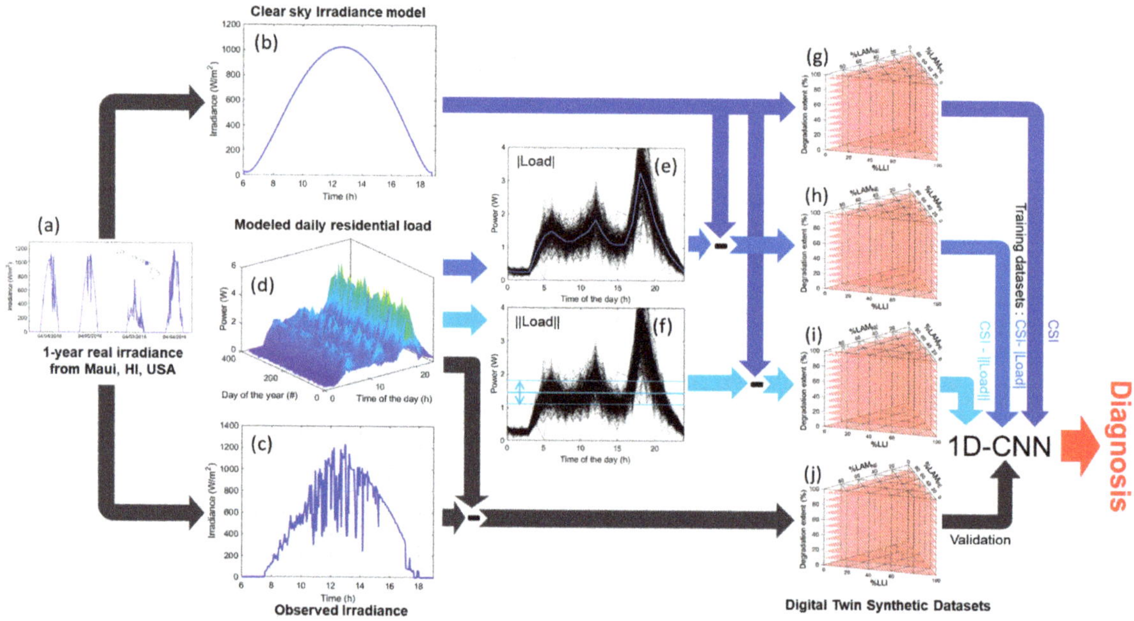

Figure 1. Schematic representation of the dataset generated in this work to assess the impact of additional residential usage with (**a**) an example of the observed irradiance for four days within the one-year dataset. (**b**) The calculated clear sky irradiance of one randomly selected day, for which (**c**) presents the observed irradiance. (**d**) The simulated daily residential power usage for one year, with (**e**) presenting the average value and (**f**) the average of the average along with ±20% variations. (**g–i**) Illustrates the three different training datasets and (**j**) the one used for validation.

2. Materials and Methods

2.1. PV Data Collection

The PV data used in this work were gathered throughout the year 2017 at a test site located at the Maui Economic Development Board (MEDB) office on the southwestern coast of the island of Maui, Hawai'i, USA. The panels were oriented at a 20° tilt with a 197° N azimuth and instrumented with a Kipp & Zonen SMP21-A (OTT HydroMet B.V., Delft, The Netherlands) secondary standard pyranometer installed at plane of array. The data were collected at 1 s intervals and averaged to 1 min. Examples of the data are provided in Figure 1a,c.

2.2. Clear Sky Irradiance Model (CSM)

The CSM used in this work is the same that was used in our previous work [23–25] implemented in MATLAB (version 2024) based on the work by Ineichen and Perez [47] for a horizontal surface with modifications to estimate clear sky irradiance on a tilted surface [48] and adding a ground-reflected irradiance source [49]. Examples of the simulated data are presented in Figure 1b. The clear sky percentage (cs%) was calculated as the percentage of common observations between the real irradiance to the output of the CSM.

2.3. Residential Usage Simulations

Residential power consumption was simulated using the HOMER Pro software (version 3.16.2), assuming a residential system with a 5.85 kW PV system associated with a 14 kWh BESS modeled with a 'load following' approach where PV generation is prioritized for household consumption, and surplus energy is used to charge the battery. Daily energy consumption was modeled with typical residential usage values. This modeling approach allowed for the replication of realistic usage profiles under various irradiance conditions. The result of the simulations is presented in Figure 1d.

2.4. Battery Emulation

The battery electrochemical behavior emulation was described in detail in the first publication of this project [23], and it will not be repeated here. The battery digital twin was built in HNEI's 'alawa mechanistic battery model [45,50] implemented in MATLAB using half-cell data from a commercial cell comprising a graphite (G) negative electrode (NE) and an NMC positive electrode (PE) with a 1:1:1 stoichiometry. The model parameters were obtained by fitting experimental data at different rates [23]. Inhomogeneities were emulated by increasing the rate degradation factors (RDF) by up to 50% to match what was observed in the literature [51]. To account for cell-to-cell variations and to avoid any overfitting error, each simulation was performed with parameters randomly varied by $\pm 1\%$ to be in the same range as the observed variations in commercial cells [52].

2.5. Synthetic Data Generation

As described in earlier work [15,17], each thermodynamic degradation can be characterized by how much lithium, PE, and NE were lost. Therefore, scanning every combination of these three degradation modes allows for the simulation of every possible degradation. Four different experiments were simulated in this work to investigate different usages, variations around the mean usage, the impact of inhomogeneities, and both at the same time. For the first one, illustrated in Figure 1g–i, the same resolution step as in our previous work (0.025%) was used for the degradation paths with 1% degradation steps. This corresponds to 861 unique triplets of the loss of lithium inventory (LLI), LAM_{PE}, and LAM_{NE} with 50 simulations per triplet (from 0 to 50% degradation for each mode with 1% increments), resulting in around 45,000 unique degradation per simulation for training. For validation, Figure 1j, the resolution step was doubled to 0.05%, which lowered the number of simulations to around 11,000 (231 unique triplets). Simulations typically lasted less than 4 days on a Windows 11 desktop with an Intel i9-14900 2 GHz processor and 64 GB of RAM. Since the second and third experiments consisted of more duty cycles (seven for the impact of variations around the mean usage and 16 for the impact of inhomogeneities), the resolution step was increased to 0.0375% for the training and 0.075% for the validation, thus keeping the 2:1 ratio between training and validation but reducing the number of triplets to 378 and 105, respectively. In addition, simulations were only undertaken to 25% maximum degradation with a 1.5% step. Overall, this reduced the number of simulations by one order of magnitude for the validation with calculations lasting less than 24 h per dataset while still providing a good coverage of all possible degradation paths (Figure A1a,b). Finally, the final simulations were undertaken to compare the initial approach to the one accounting for the impact of additional usage and battery packs. For these simulations, the resolution step was set at 0.02% for the training and 0.05% for the validation. This was undertaken to limit the overlap between the training and validation datasets (Figure A1c), and the simulations took around 4.5 days.

As proposed in [23,29], the non-constant current duty cycles were calculated by simulating 100 different rates for each SOH. The duty cycles were constructed step by step by using the voltage-rate couple matching the power request at the current SOC for each step. Since the current was not constant, capacity and time are decorrelated and the derivatives for both versus voltage (dQ/dV and dt/dV) were used for diagnosis.

2.6. 1D-CNN Implementation

This work used the same convolutional neural network (CNN) as in [23–25], as it was shown to offer a good compromise between accuracy and ease of implementation [23]. It was originally developed by Kim et al. [18] and was implemented in TensorFlow (https://www.tensorflow.org/, accessed on 13 April 2025) [53] with 2 CNN-1D layers with 32 neurons each and 3 fully connected layers with 128, 64, and 3 neurons each. The batch size, learning rate and the number of epochs were fixed to 64, 0.001, and 25, respectively. Similarly to what was implemented in [23], the algorithm was trained, validated, and tested on both voltage vs. capacity and voltage vs. time curves. They will be referred to as Q-based and t-based diagnosis, respectively.

2.7. Statistical Metrics

The RMSE was defined as follows:

$$RMSE = \sqrt{\sum_{i=1}^{n} \frac{(y_i - x_i)^2}{n}},$$

with y_i being the prediction, x_i the true value, and n the number of data points.

3. Results

3.1. Observed vs. Clear Sky Irradiance

Figure 2a presents the distribution of the OIs and CSIs throughout 2017 at the MEDB site. The OIs varied between 67 and 651 W/m^2 with a mean of 505 W/m^2. Most of the OIs were above 400 W/m^2 with a median at 532 W/m^2 with a 104 W/m^2 interquartile range. Unsurprisingly, the calculated average CSIs are higher with an average (and median) of 630 W/m^2 with a minimum at 595 and a maximum at 666 W/m^2. The distribution is less dispersed than that of the OI (interquartile range of 45 W/m^2), and is bimodal with a mode around 600 W/m^2 and another one around 650 W/m^2. This bimodality can be explained by the orientation and tilt of the panels [23]. Looking at the cs%, more than 80 out of 365 days had less than 5% clear sky, whereas only 16 had 80% or more clear sky. Overall, the average and median values were around 33%, with 22% of the days having 50% or more clear sky.

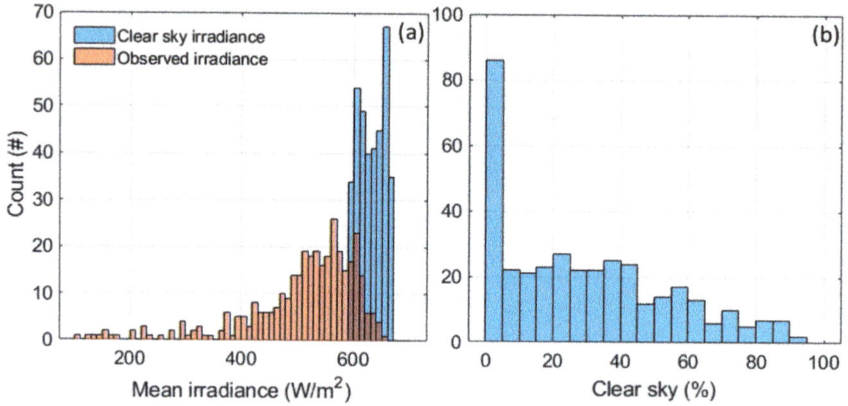

Figure 2. (**a**) Distribution of the observed vs. clear sky mean irradiances through 2017 and (**b**) associated clear sky percentages.

3.2. Residential Load Simulations

Figure 3a presents an example of the residential load simulation obtained from the HOMER software with a different mix of power units to match the demand throughout the day. The day started with some grid usage early in the morning, followed by power generation from the PV system as solar radiation increased from sunrise around 7 a.m.

Once the PV system generated more power than required, around 8 a.m., the excess power was used to charge the battery, and it was curtailed once the battery was fully charged around 2 p.m. Past sunset, the battery supplied the necessary load until it was depleted around 9 p.m., after which the grid took over. Figure 3b highlights the differences between the power that would have been gathered from the clear sky irradiance (used for training in our previous work [23–25]), the power actually gathered from the PV system (used for validation in our previous work [23–25]), and the actual power used by the battery. This illustrates the necessity to take this additional usage into account. For the algorithm validation, this could be achieved by using the power obtained from the OI minus the variable daily simulated residential load (Figure 3c). It is more complicated, however, for the training data, as the daily usage is variable and, for a deployed application, might not be known beforehand. A possible solution is to subtract the average daily load (| Load |) from the clear sky power (Figure 3d), but this still implies some knowledge about the usage. Another possibility could be to subtract a constant value that could be estimated to be close to the average of the average load (| | Load | |), as shown in Figure 3d. The results of these subtractions are presented, with the dotted curves in Figure 3b showcasing much closer values to the actual charging power for the BESS.

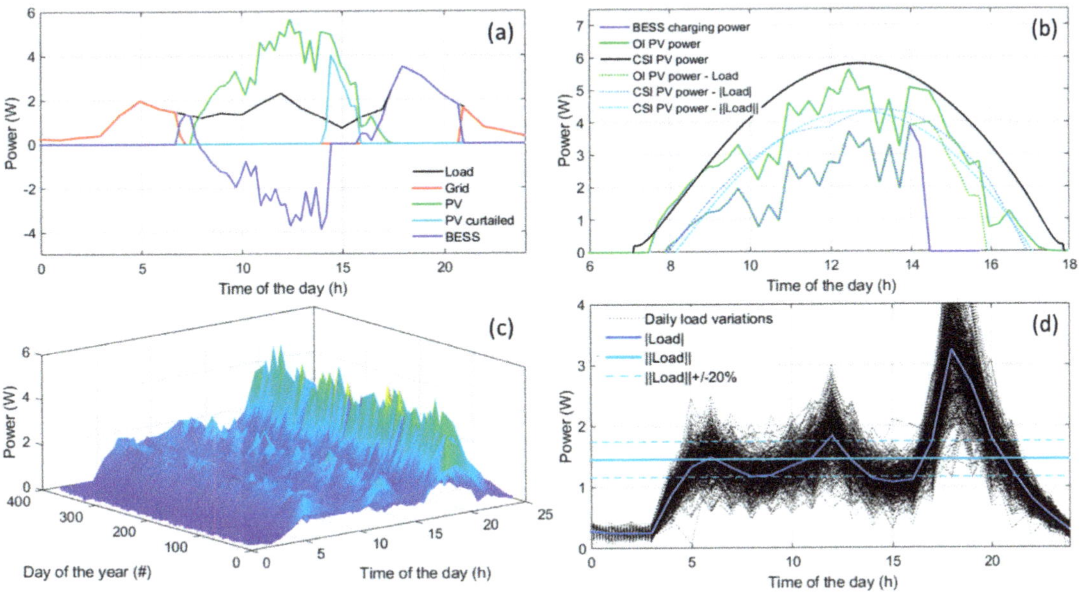

Figure 3. (**a**) Example of the residential load simulations for 1 January 2017, showcasing the mix of origins for required power between the PV system, the grid, and the battery. (**b**) Comparison of the power generated from the clear sky irradiance, the observed irradiance, and the one used to charge the battery on 1 January with or without adjustments to accommodate additional loads. (**c**) Variability of the simulated residential load as a function of the day of the year with (**d**) the associated average usage, and the average of the average usage ± 20%.

3.3. Impact of Residential Usage

Figure 4 presents the average RMSEs for the 11,000 considered degradations over 365 days as a function of the clear sky percentage in 10% increments. The results are presented as box plots where the size of the box represents 50% of the data, with the median represented by the line. The whiskers represent a 3σ deviation from the median and the circles represent the outliers. The top row presents the results for up to 50% degradation and the bottom one up to 25%. The left column shows the results from a Q-based diagnosis and the right column shows the results from a t-based one.

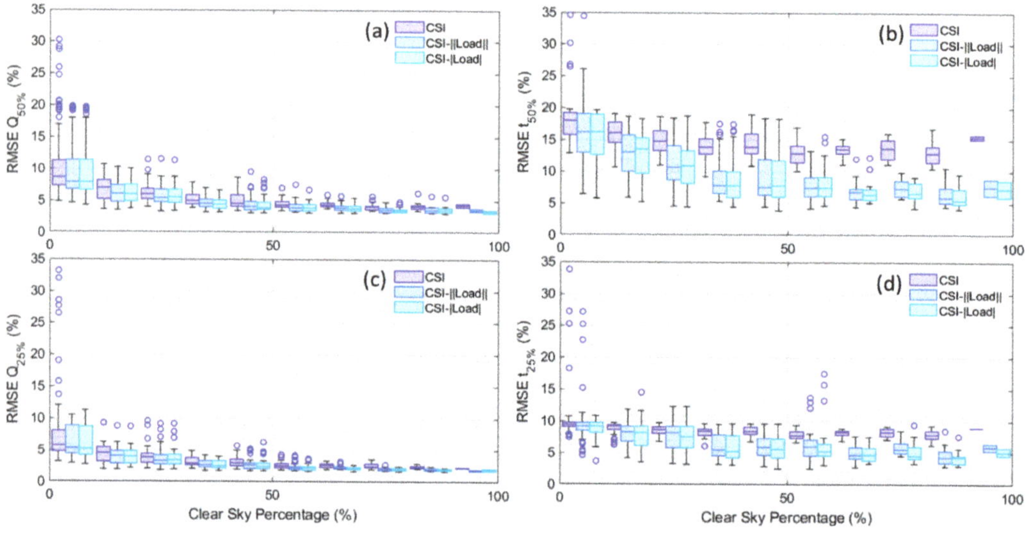

Figure 4. Evolution of the RMSE for diagnosis for (**a**) up to 50% capacity loss based on capacity, (**b**) up to 50% capacity loss based on time, (**c**) up to 25% capacity loss based on capacity, and (**d**) up to 50% capacity loss based on time.

From a Q-based standpoint, the type of dataset used for training does not seem to have a major impact as the results are close between the CSI, CSI-|Load|, and CSI-||Load|| datasets. Nonetheless, there seems to be consistent ordering, with the RMSE for the CSI only being slightly higher than that for CSI-||Load|| and then CSI-|Load|. This ordering is much more visible when the t-based diagnosis is considered (right column), where there is a clear drop in RMSE between the CSI-trained data and the data trained with CSI minus a load, which are close but still with an advantage for CSI-|Load|.

Figure 5a,b present the evolution of the average RMSE for both the Q- and t-based diagnosis as a function of the training dataset for all days (black lines) and for days with a cs% > 50% (blue lines). Figure 5a,b confirm the observations from Figure 4 with the worst performance for the CSI-trained algorithms both for the Q- and t-based diagnosis with a bigger difference for the t-based diagnosis. The average performance for the CSI-|Load| and CSI-||Load|| is extremely similar and likely well within the margin of error for our approach. Being around 2%, this performance is also just slightly higher than the 1.75% obtained without considering additional usage [23,24]. Figure 5c displays the impact of moving the value of ||Load|| by ±20% for days with more than 50% clear sky. These variations have very little effect when days with >50 cs% are considered and only a mild effect on the t-based diagnosis when all days are considered with the lowest RMSEs for the −20% dataset.

3.4. Impact of Pack Imbalance/Inhomogeneities

Figure 6a showcases the impact of increasing the RDF on the derivative of the simulated electrochemical response of the cell at C/4. More information and details on the significance of changing the RDF can be found in previous works [29,45,46]. The impact is especially visible on the local minima around 3.8 V that become progressively filled when the RDF increases by 50%. Additional effects can be seen between the low voltage and the main peak, where the two small graphitic peaks are no longer visible because of the broadening of the peaks. These voltage changes might affect the algorithm's ability to recognize the different degradations. Figure 6b,c presents the average RMSEs associated with simulations for different levels of inhomogeneities both for the training and validation datasets. For the training datasets, the CSI dataset was used for generation with either the initial cell, the initial cell plus 25% RDF, or the initial cell plus 50% RDF. The validation dataset was generated from the OI minus the daily residential load using a cell with either 0%, 12.5%, 25%, 37.5%, or 50% increased RDF.

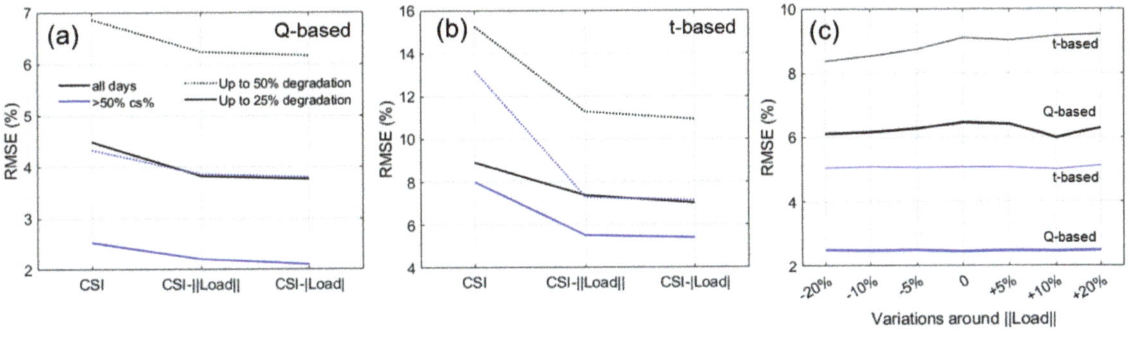

Figure 5. Effect of the training dataset on the average RMSE for (**a**) the Q-based diagnosis and (**b**) the t-based diagnosis. (**c**) Effect of variations around | | Load | | on the average RMSE.

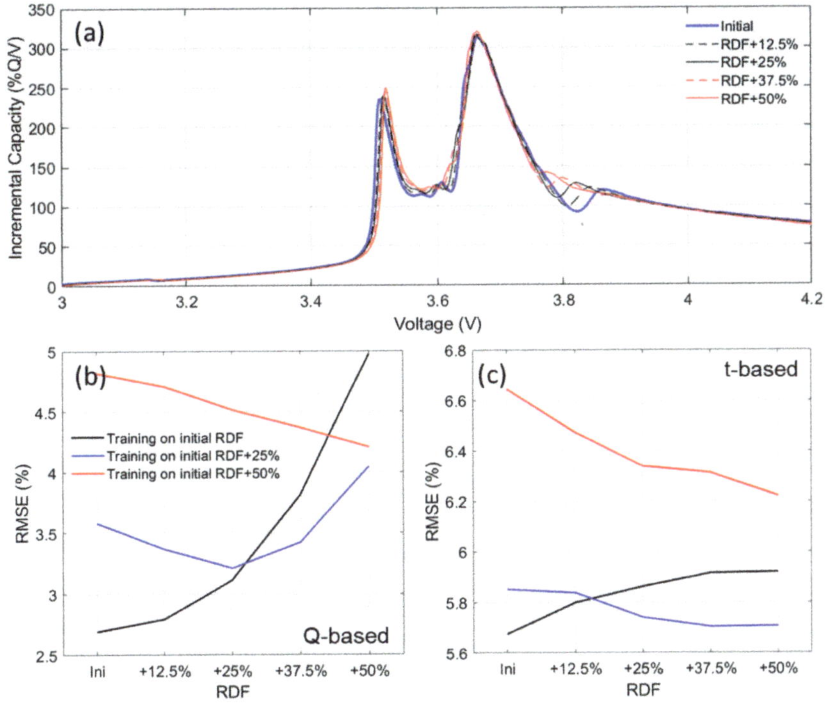

Figure 6. (**a**) Effect of increasing the RDF on the cell electrochemical response. Effect of the amount of inhomogeneities on the average RMSE for (**b**) the Q-based diagnosis and (**c**) the t-based diagnosis.

The error associated with an increasing RDF for the validation cycles increases for the algorithm trained on the initial cell and decreases for the algorithms trained with the initial cell plus 50% RDF. For the algorithms trained with the initial cell plus 25% RDF, a bell shape curve is observed for the Q-based diagnosis RMSEs, with a minimum for the validation data with +25% RDF. This minimum was around 1% above the homogeneous dataset average RMSEs. The bell shape is not observed for the t-based diagnosis, where the highest error was observed for lower RDFs.

4. Discussion

Looking first into the impact of additional usage on the cells, it must be noted that the HOMER-generated residential load profiles used in this work relied on empirical consumption patterns, reflecting typical household behavior. While these profiles smoothed over stochastic variations, such as those caused by individual occupant activities, we believe this approximation was suitable for the synthetic data generated in this work. Our simulations indicated only a small impact of the additional usage on the Q-based diagnosis between the unaccommodated simulations and the accommodated ones, whether by subtracting the average load or the average of the average load. Indeed, from Figure 5a, it can be seen that the average RMSE decreased by less than 1% in the all-days scenario but by less than 0.5% if only the days with 50% or more clear sky are considered, independently of the extent of the degradation. For the t-based diagnosis, the impact is much larger, especially for the larger capacity losses where 3 to 5% gain in RMSEs was calculated. As discussed in [23], the difference between the Q-based and t-based diagnosis is not surprising. Indeed, since the charge of the cells is overall rather slow, lasting more than 10 h, variations in current do not significantly impact the voltage nor the exchanged capacity, but since these variations will affect charging times, the t-based diagnosis will be affected. Since the diagnosis accuracy in the cases of training using |load|, ||load||, and ±20% around ||load|| was rather constant (Figure 5b,c), this same impact of a slow charge seems to remove the necessity of knowing the exact usage associated with each battery to be able to accommodate their effect, which is good news for applicability.

Looking into the impact of transitioning from a single-cell model to a pack model (Figure 6b,c), it appeared to be larger than that for the training datasets, with variations of more than 2% in the average RMSE reported depending on the conditions. Training on a single cell will work well for packs with little inhomogeneity but the RMSE will quickly increase with an increase in RDF, especially for Q-based diagnosis. The opposite is true when training on a model with significant inhomogeneities. The best compromise seemed to be training on a cell with mild inhomogeneities (+25% RDF in our case). In such cases, a bell-shaped curve was observed for Q-based RMSEs with less than 1% variation between the minimum and the maximum.

To investigate the impact of combining both accommodations, the last experiment consisted first of a dataset trained on the CSI without any accommodation and second a dataset trained with the CSI-||load|| with also +25% on the RDF to account for the impact of the additional usage and of inhomogeneities. The validation was performed both on a dataset using the OI minus the residential usage and a random value for the RDF increase between 0 and 50% for each day of the year. The results of these simulations are presented in Figure 7. Unsurprisingly, the t-based diagnosis benefited the most from both accommodations with an improvement in the average RMSE between 1.75 and 5.1% depending on the conditions. The improvement was much more limited for the Q-based diagnosis, where it stayed below 1%. Nonetheless, a closer investigation of the average RMSEs as a function of the type of degradation (Figure 7c) showcased some improvement, with lower RMSEs in the regions where the algorithm struggled the most. Looking at the Q-based diagnosis up to 50% (first column), and without any accommodations, the diagnosis error is the highest for degradation paths with around 50% LAM_{NE} (yellow area on top left triangle indicating RMSEs around 10%). After accommodation, the yellow area disappears, and the average error is much more consistent across the entire degradation paths. The same can be observed for both t-diagnosis (3rd and 4th columns), where the color distribution is much more homogeneous after accommodation, indicating an overall better performance of the algorithm. This is promising for deployment, as this indicates that the error is less dependent on the actual degradation and that, although the best diagnoses were not improved much, the worst ones were.

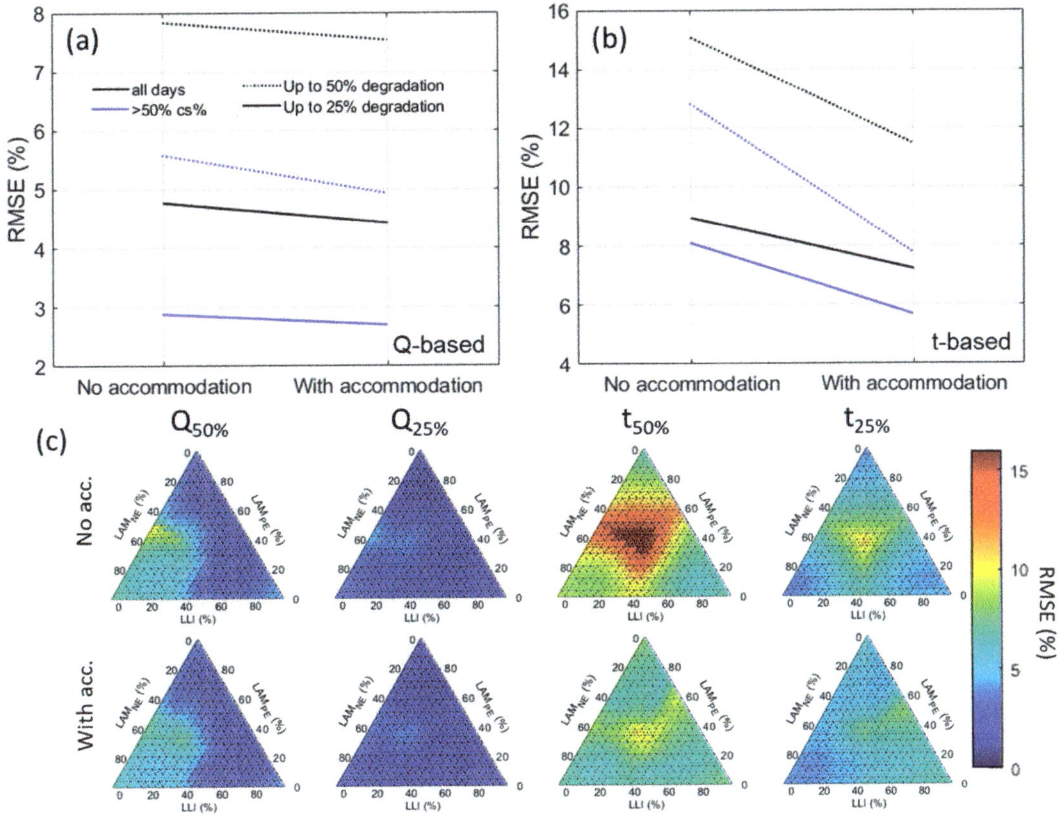

Figure 7. Effect of the accommodation on the average RMSE for (**a**) the Q-based diagnosis and (**b**) the t-based diagnosis. (**c**) Diagnosis accuracy for days with >50% cs% as a function of the degradation path for simulation without accommodation (**top row**) and with accommodation (**bottom row**).

Overall, the accommodation of additional usage and battery packs only increased the average RMSEs for diagnosis by 1% compared to the ideal scenario (2.75% vs. 1.75% for training and validation on single cells without additional usage [23,24]). Looking to further improve the diagnosability, possible improvements could include better emulation of PV systems. In the example given, this could be accomplished by using a different model integrating maximum power point tracking [54,55]. In addition, the addition of forecasting for the global horizontal irradiance [56–58] could help to select the best days to perform the diagnosis. Finally, a wide variety of voltage-based diagnosis algorithms have been proposed in the literature [59,60], and some could potentially improve on the algorithm used in this work. Since this work was focused on data generation, scoping different types of algorithms was out of scope.

5. Conclusions

Following on from our previous work where we proposed a proof-of-concept approach for the diagnosis of PV-connected batteries without the need for maintenance cycles, this work addressed the two main limitations of the initial approach, where only single cells without any additional usage on the battery were considered. This work showed that, for capacity-based diagnosis, these limitations have a limited impact of the diagnosis accuracy and that they can be accommodated without adding much complexity to the approach by subtracting the average usage from the clear sky irradiance and by accounting for a medium level of inhomogeneities. These accommodations, while of limited impact for the average capacity-based diagnosis, with around a 1% increase in the average RMSE for the best-case figure, significantly improved the worst Q-based diagnoses and all the time-based diagnoses by up to 5% depending on the conditions.

Overall, this study reinforced the potential of synthetic data generation and training to enable Big-Data studies of the behavior of deployed systems experiencing sporadic usage.

Author Contributions: Conceptualization, M.D.; methodology, M.D. and S.S.; software, M.D. and S.S.; experimentation, F.Y.; validation, F.Y. and M.D.; formal analysis, F.Y.; investigation, M.D.; resources, M.D.; data curation, M.D.; writing—original draft preparation, M.D.; writing—review and editing, F.Y., S.S. and M.D.; visualization, M.D.; supervision, M.D.; project administration, M.D.; funding acquisition, M.D. All authors have read and agreed to the published version of the manuscript.

Funding: This research was funded by ONR grand number # N00014-21-1-2250. The authors are thankful to ONR for funding the deployment and monitoring of the PV system used in this work with the help of HNEI's Severine Busquet, Jonathan Kobayashi, and Richard Rocheleau.

Data Availability Statement: Given the low calculation time necessary for each simulation and the amount of data generated (well over 40 GB), the synthetic data used in this work were not saved. The data used for this work will be resimulated upon reasonable request at the discretion of the authors.

Acknowledgments: The authors gratefully acknowledge the contributions of Nahuel Costa and Dax Matthews for their work at the beginning of this project.

Conflicts of Interest: The authors declare no conflicts of interest.

Abbreviations

The following abbreviations are used in this manuscript:

BESS	Battery energy storage system
CNN	Convolutional neural network
CSI	Clear sky irradiance
CSM	Clear sky irradiance model
LAM	Loss of active material
LLI	Loss of lithium inventory
MEDB	Maui Economic Development Board
NE	Negative electrode
NMC	Nickel manganese cobalt oxide
OI	Observed irradiance
PE	Positive electrode
RDF	Rate degradation factor
RMSE	Root mean square error
SOC	State of charge
SOH	State of health

Appendix A

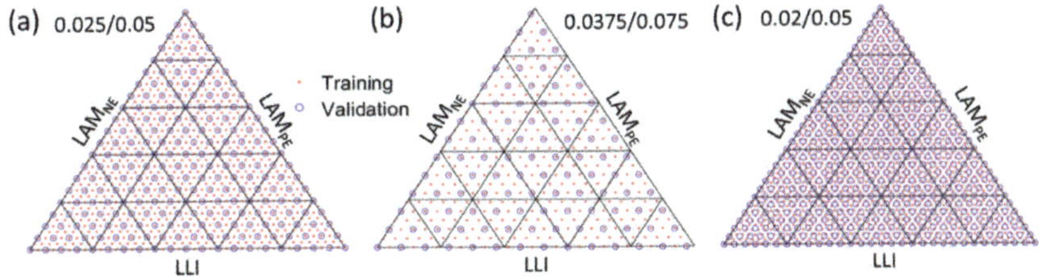

Figure A1. Degradation path resolution for the (**a**) 0.025/0.05, (**b**) 0.0375/0.075, and (**c**) 0.02/0.05 simulations.

References

1. Anukoolthamchote, P.C.; Assané, D.; Konan, D.E. Net electricity load profiles: Shape and variability considering customer-mix at transformers on the island of Oahu, Hawai'i. *Energy Policy* **2020**, *147*, 111732. [CrossRef]
2. Covelli, D.; Virgüez, E.; Caldeira, K.; Lewis, N.S. Oahu as a case study for island electricity systems relying on wind and solar generation instead of imported petroleum fuels. *Appl. Energy* **2024**, *375*, 124054. [CrossRef]
3. Che, Y.; Hu, X.; Lin, X.; Guo, J.; Teodorescu, R. Health prognostics for lithium-ion batteries: Mechanisms, methods, and prospects. *Energy Environ. Sci.* **2023**, *16*, 338–371. [CrossRef]
4. Vasta, E.; Scimone, T.; Nobile, G.; Eberhardt, O.; Dugo, D.; De Benedetti, M.M.; Lanuzza, L.; Scarcella, G.; Patanè, L.; Arena, P.; et al. Models for Battery Health Assessment: A Comparative Evaluation. *Energies* **2023**, *16*, 632. [CrossRef]
5. Barrett, D.H.; Haruna, A. Artificial intelligence and machine learning for targeted energy storage solutions. *Curr. Opin. Electrochem.* **2020**, *21*, 160–166. [CrossRef]
6. Cui, Z.; Wang, L.; Li, Q.; Wang, K. A comprehensive review on the state of charge estimation for lithium-ion battery based on neural network. *Int. J. Energ. Res.* **2021**, *46*, 5423–5440. [CrossRef]
7. Sharma, P.; Bora, B.J. A Review of Modern Machine Learning Techniques in the Prediction of Remaining Useful Life of Lithium-Ion Batteries. *Batteries* **2022**, *9*, 13. [CrossRef]
8. Rauf, H.; Khalid, M.; Arshad, N. Machine learning in state of health and remaining useful life estimation: Theoretical and technological development in battery degradation modelling. *Renew. Sustain. Energy Rev.* **2022**, *156*, 111903. [CrossRef]
9. Na, H.S.; Numan-Al-Mobin, A.M. Machine learning approaches to estimate the health state of next-generation energy storage. In *Green Sustainable Process for Chemical and Environmental Engineering and Science*; Elsevier: Amsterdam, The Netherlands, 2023; pp. 343–363.
10. Severson, K.A.; Attia, P.M.; Jin, N.; Perkins, N.; Jiang, B.; Yang, Z.; Chen, M.H.; Aykol, M.; Herring, P.K.; Fraggedakis, D.; et al. Data-driven prediction of battery cycle life before capacity degradation. *Nat. Energy* **2019**, *4*, 383–391. [CrossRef]
11. Röder, F.; Ramasubramanian, S. A Review and Perspective on Path Dependency in Batteries. *Energy Technol.* **2022**, *10*, 2200627. [CrossRef]
12. Srinivasan, V.; Newman, J. Existence of Path-Dependence in the $LiFePO_4$ Electrode. *Electrochem. Solid-State Lett.* **2006**, *9*, A110. [CrossRef]
13. Gering, K.L.; Sazhin, S.V.; Jamison, D.K.; Michelbacher, C.J.; Liaw, B.Y.; Dubarry, M.; Cugnet, M. Investigation of path dependence in commercial lithium-ion cells chosen for plug-in hybrid vehicle duty cycle protocols. *J. Power Sources* **2011**, *196*, 3395–3403. [CrossRef]
14. Figgener, J.; Bors, J.; Kuipers, M.; Hildenbrand, F.; Junker, M.; Koltermann, L.; Woerner, P.; Mennekes, M.; Haberschusz, D.; Kairies, K.-P.; et al. Degradation mode estimation using reconstructed open circuit voltage curves from multi-year home storage field data. *ArXiv* **2025**, arXiv:2411.08025. [CrossRef]
15. Dubarry, M.; Berecibar, M.; Devie, A.; Anseán, D.; Omar, N.; Villarreal, I. State of health battery estimator enabling degradation diagnosis: Model and algorithm description. *J. Power Sources* **2017**, *360*, 59–69. [CrossRef]
16. Dubarry, M.; Beck, D. Analysis of Synthetic Voltage vs. Capacity Datasets for Big Data Li-ion Diagnosis and Prognosis. *Energies* **2021**, *14*, 2371. [CrossRef]
17. Dubarry, M.; Beck, D. Big data training data for artificial intelligence-based Li-ion diagnosis and prognosis. *J. Power Sources* **2020**, *479*, 228806. [CrossRef]
18. Kim, S.; Yi, Z.; Chen, B.-R.; Tanim, T.R.; Dufek, E.J. Rapid failure mode classification and quantification in batteries: A deep learning modeling framework. *Energy Storage Mater.* **2022**, *45*, 1002–1011. [CrossRef]

19. Mayilvahanan, K.S.; Takeuchi, K.J.; Takeuchi, E.S.; Marschilok, A.C.; West, A.C. Supervised Learning of Synthetic Big Data for Li-Ion Battery Degradation Diagnosis. *Batter. Supercaps* **2021**, *5*, e202100166. [CrossRef]
20. Costa, N.; Sánchez, L.; Anseán, D.; Dubarry, M. Li-ion battery degradation modes diagnosis via Convolutional Neural Networks. *J. Energy Storage* **2022**, *55*, 105558. [CrossRef]
21. Kim, S.; Jung, H.; Lee, M.; Choi, Y.Y.; Choi, J.-I. Model-free reconstruction of capacity degradation trajectory of lithium-ion batteries using early cycle data. *eTransportation* **2023**, *17*, 100243. [CrossRef]
22. Ruan, H.; Chen, J.; Ai, W.; Wu, B. Generalised diagnostic framework for rapid battery degradation quantification with deep learning. *Energy AI* **2022**, *9*, 100158. [CrossRef]
23. Dubarry, M.; Costa, N.; Matthews, D. Data-driven direct diagnosis of Li-ion batteries connected to photovoltaics. *Nat. Commun.* **2023**, *14*, 3138. [CrossRef]
24. Dubarry, M.; Yasir, F.; Costa, N.; Matthews, D. Data-Driven Diagnosis of PV-Connected Batteries: Analysis of Two Years of Observed Irradiance. *Batteries* **2023**, *9*, 395. [CrossRef]
25. Dubarry, M.; Yasir, F. Big Data for the Diagnosis and Prognosis of Deployed Energy Storage Systems. In Proceedings of the 2024 IEEE Electrical Energy Storage Application and Technologies Conference (EESAT), San Diego, CA, USA, 29–30 January 2024; pp. 1–5.
26. Li, R.; O'Kane, S.; Huang, J.; Marinescu, M.; Offer, G.J. A million cycles in a day: Enabling high-throughput computing of lithium-ion battery degradation with physics-based models. *J. Power Sources* **2024**, *598*, 234184. [CrossRef]
27. Hofmann, T.; Hamar, J.; Mager, B.; Erhard, S.; Schmidt, J.P. Transfer learning from synthetic data for open-circuit voltage curve reconstruction and state of health estimation of lithium-ion batteries from partial charging segments. *Energy AI* **2024**, *17*, 100382. [CrossRef]
28. Ruan, H.; Kirkaldy, N.; Offer, G.J.; Wu, B. Diagnosing health in composite battery electrodes with explainable deep learning and partial charging data. *Energy AI* **2024**, *16*, 100352. [CrossRef]
29. Dubarry, M.; Beck, D. Perspective on Mechanistic Modeling of Li-Ion Batteries. *Acc. Mater. Res.* **2022**, *3*, 843–853. [CrossRef]
30. Dubarry, M.; Vuillaume, N.; Liaw, B.Y. From single cell model to battery pack simulation for Li-ion batteries. *J. Power Sources* **2009**, *186*, 500–507. [CrossRef]
31. Kim, J.; Cho, B.H. Stable Configuration of a Li-Ion Series Battery Pack Based on a Screening Process for Improved Voltage/SOC Balancing. *IEEE Trans. Power Electron.* **2012**, *17*, 411–424. [CrossRef]
32. Jiang, Y.; Jiang, J.; Zhang, C.; Zhang, W.; Gao, Y.; Guo, Q. Recognition of battery aging variations for $LiFePO_4$ batteries in 2nd use applications combining incremental capacity analysis and statistical approaches. *J. Power Sources* **2017**, *360*, 180–188. [CrossRef]
33. Tanim, T.R.; Dufek, E.J.; Sazhin, S.V. Challenges and needs for system-level electrochemical lithium-ion battery management and diagnostics. *MRS Bull.* **2021**, *46*, 420–428. [CrossRef]
34. Xu, Z.; Wang, J.; Lund, P.D.; Zhang, Y. Estimation and prediction of state of health of electric vehicle batteries using discrete incremental capacity analysis based on real driving data. *Energy* **2021**, *225*, 120160. [CrossRef]
35. Chang, L.; Wang, C.; Zhang, C.; Xiao, L.; Cui, N.; Li, H.; Qiu, J. A novel fast capacity estimation method based on current curves of parallel-connected cells for retired lithium-ion batteries in second-use applications. *J. Power Sources* **2020**, *459*, 227901. [CrossRef]
36. Krupp, A.; Ferg, E.; Schuldt, F.; Derendorf, K.; Agert, C. Incremental Capacity Analysis as a State of Health Estimation Method for Lithium-Ion Battery Modules with Series-Connected Cells. *Batteries* **2020**, *7*, 2. [CrossRef]
37. Jiang, T.; Sun, J.; Wang, T.; Tang, Y.; Chen, S.; Qiu, S.; Liu, X.; Lu, S.; Wu, X. Sorting and grouping optimization method for second-use batteries considering aging mechanism. *J. Energy Storage* **2021**, *44*, 103264. [CrossRef]
38. Wassiliadis, N.; Steinsträter, M.; Schreiber, M.; Rosner, P.; Nicoletti, L.; Schmid, F.; Ank, M.; Teichert, O.; Wildfeuer, L.; Schneider, J.; et al. Quantifying the state of the art of electric powertrains in battery electric vehicles: Range, efficiency, and lifetime from component to system level of the Volkswagen ID.3. *eTransportation* **2022**, *12*, 100167. [CrossRef]
39. Singh, A.; Lodge, A.; Li, Y.; Widanage, W.D.; Barai, A. A new method to perform Lithium-ion battery pack fault diagnostics—Part 1: Algorithm development and its performance analysis. *Energy Rep.* **2023**, *10*, 4474–4490. [CrossRef]
40. Bilfinger, P.; Rosner, P.; Schreiber, M.; Kröger, T.; Gamra, K.A.; Ank, M.; Wassiliadis, N.; Dietermann, B.; Lienkamp, M. Battery pack diagnostics for electric vehicles: Transfer of differential voltage and incremental capacity analysis from cell to vehicle level. *eTransportation* **2024**, *22*, 100356. [CrossRef]
41. Dubarry, M.; Tun, M.; Baure, G.; Matsuura, M.; Rocheleau, R.E. Battery Durability and Reliability under Electric Utility Grid Operations: Analysis of On-Site Reference Tests. *Electronics* **2021**, *10*, 1593. [CrossRef]
42. Jocher, P.; Roehrer, F.; Rehm, M.; Idrizi, T.; Himmelreich, A.; Jossen, A. Scaling from cell to system: Comparing Lithium-ion and Sodium-ion technologies regarding inhomogeneous resistance and temperature in parallel configuration by sensitivity factors. *J. Energy Storage* **2024**, *98*, 112931. [CrossRef]
43. Rosenberger, N.; Rosner, P.; Bilfinger, P.; Schöberl, J.; Teichert, O.; Schneider, J.; Abo Gamra, K.; Allgäuer, C.; Dietermann, B.; Schreiber, M.; et al. Quantifying the State of the Art of Electric Powertrains in Battery Electric Vehicles: Comprehensive Analysis of the Tesla Model 3 on the Vehicle Level. *World Electr. Veh. J.* **2024**, *15*, 268. [CrossRef]

44. Dubarry, M.; Beck, D. Investigation of the impact of different electrode inhomogeneities on the voltage response of Li-ion batteries. *Cell Rep. Phys. Sci.* **2024**, 102138, in press. [CrossRef]
45. Dubarry, M.; Truchot, C.; Liaw, B.Y. Synthesize battery degradation modes via a diagnostic and prognostic model. *J. Power Sources* **2012**, *219*, 204–216. [CrossRef]
46. Schindler, S.; Baure, G.; Danzer, M.A.; Dubarry, M. Kinetics accommodation in Li-ion mechanistic modeling. *J. Power Sources* **2019**, *440*, 227117. [CrossRef]
47. Ineichen, P.; Perez, R. A new airmass independent formulation for the Linke turbidity coefficient. *Sol. Energy* **2002**, *73*, 151–157. [CrossRef]
48. Liu, B.Y.H.; Jordan, R.C. The interrelationship and characteristic distribution of direct, diffuse and total solar radiation. *Sol. Energy* **1960**, *4*, 1–19. [CrossRef]
49. Loutzenhiser, P.G.; Manz, H.; Felsmann, C.; Strachan, P.A.; Frank, T.; Maxwell, G.M. Empirical validation of models to compute solar irradiance on inclined surfaces for building energy simulation. *Sol. Energy* **2007**, *81*, 254–267. [CrossRef]
50. HNEI. Alawa Central. Available online: https://www.hnei.hawaii.edu/alawa (accessed on 13 April 2025).
51. Sieg, J.; Storch, M.; Fath, J.; Nuhic, A.; Bandlow, J.; Spier, B.; Sauer, D.U. Local degradation and differential voltage analysis of aged lithium-ion pouch cells. *J. Energy Storage* **2020**, *30*, 101582. [CrossRef]
52. Devie, A.; Dubarry, M. Durability and Reliability of Electric Vehicle Batteries under Electric Utility Grid Operations. Part 1: Cell-to-Cell Variations and Preliminary Testing. *Batteries* **2016**, *2*, 28. [CrossRef]
53. Abadi, M.; Barham, P.; Chen, J.; Chen, Z.; Davis, A.; Dean, J.; Devin, M.; Ghemawat, S.; Irving, G.; Isard, M.; et al. TensorFlow: A system for large-scale machine learning. In Proceedings of the 12th USENIX Symposium on Operating Systems Design and Implementation (OSDI '16), Savannah, GA, USA, 2–4 November 2016.
54. Harrison, A.; Alombah, N.H.; Kamel, S.; Kotb, H.; Ghoneim, S.S.M.; El Myasse, I. A Novel MPPT-Based Solar Irradiance Estimator: Integration of a Hybrid Incremental Conductance Integral Backstepping Algorithm for PV Systems with Experimental Validation. *Eng. Proc.* **2023**, *56*, 262. [CrossRef]
55. Harrison, A.; Alombah, N.H.; Kamel, S.; Ghoneim, S.S.M.; El Myasse, I.; Kotb, H. Towards a Simple and Efficient Implementation of Solar Photovoltaic Emulator: An Explicit PV Model Based Approach. *Eng. Proc.* **2023**, *56*, 261. [CrossRef]
56. Elmousaid, R.; Drioui, N.; Elgouri, R.; Agueny, H.; Adnani, Y. Accurate short-term GHI forecasting using a novel temporal convolutional network model. *e-Prime Adv. Electr. Eng. Electron. Energy* **2024**, *9*, 100667. [CrossRef]
57. Qin, C.; Srivastava, A.K.; Saber, A.Y.; Matthews, D.; Davies, K. Geometric Deep-Learning-Based Spatiotemporal Forecasting for Inverter-Based Solar Power. *IEEE Syst. J.* **2023**, *17*, 3425–3435. [CrossRef]
58. Matthews, D.K. Determination of broadband atmospheric turbidity from global irradiance or photovoltaic power data using deep neural nets. *Energy AI* **2023**, *14*, 100252. [CrossRef]
59. Vanem, E.; Wang, S. Data-driven state of health and state of safety estimation for alternative battery chemistries—A comparative review focusing on sodium-ion and LFP lithium-ion batteries. *Future Batter.* **2025**, *5*, 100033. [CrossRef]
60. De la Iglesia, D.H.; Corbacho, C.C.; Dib, J.Z.; Alonso-Secades, V.; López Rivero, A.J. Advanced Machine Learning and Deep Learning Approaches for Estimating the Remaining Life of EV Batteries—A Review. *Batteries* **2025**, *11*, 17. [CrossRef]

batteries

MDPI

Article

State-of-Health Estimation for Lithium-Ion Batteries Based on Lightweight DimConv-GFNet

Kehao Huang [1,2], Jianqiang Kang [1,2], Jing V. Wang [3], Qian Wang [3] and Oukai Wu [1,2,*]

[1] Hubei Key Laboratory of Advanced Technology for Automotive Components, Wuhan University of Technology, Wuhan 430070, China; huangkehao@whut.edu.cn (K.H.); kjqiang@whut.edu.cn (J.K.)
[2] Hubei Collaborative Innovation Center for Automotive Components Technology, Wuhan 430070, China
[3] School of Automation, Wuhan University of Technology, Wuhan 430070, China; jingvwang@whut.edu.cn (J.V.W.); qiw@whut.edu.cn (Q.W.)
* Correspondence: wuoukai@whut.edu.cn

Abstract: The accurate estimation of the state of health (SOH) is crucial for effective battery management systems. This paper proposes a deep learning model dimension-wise convolutions-globalfilter networks (DimConv-GFNet) for lithium-ion battery SOH estimation. Particularly, the DimConv-GFNet comprises the dimension-wise convolutions (DimConv), which collect the multi-scale local features from different sensor signals, and lightweight global filter networks (GFNet) to capture long-range dependencies in the Fourier frequency domain. Unlike Transformer attention architectures, GFNet utilizes spectral properties to facilitate global information exchange with a lower computational complexity. Experiments on two datasets with a total of 167 commercial LFP/graphite cells validate the effectiveness of DimConv-GFNet. Although the model shows slightly lower accuracy compared to the DimConv-Transformer baseline, it delivers competitive performance with a root mean squared error (RMSE) of 0.335%, mean absolute error (MAE) of 0.233% and a mean absolute percentage error (MAPE) of 0.230%. Remarkably, the DimConv-GFNet substantially reduces computational demands, requiring fewer than one-third of the Floating Point Operations (FLOPs) and parameters of DimConv-Transformer. These results demonstrate DimConv-GFNet strikes a good balance between accuracy and efficiency, positioning it as a promising solution for efficient and accurate SOH estimation in battery management applications.

Keywords: lithium-ion battery; state-of-health estimation; dimension-wise convolutions; Fourier transform; global filter network

1. Introduction

Lithium-ion batteries have become a critical pillar in the global transition to clean energy, owing to their high energy density, long cycle life, and low self-discharge rate. They are widely deployed in portable electronics, electric vehicles, and renewable energy storage systems [1–3], significantly enhancing energy efficiency and reducing greenhouse gas emissions, providing strong support for sustainable development. Despite their widespread adoption, lithium-ion batteries are subject to gradual degradation over their lifespan, primarily driven by inherent electrochemical side reactions and material aging [4].

The state of health (SOH), defined as the ratio of the current maximum discharge capacity to the nominal capacity, is typically utilized to assess the battery degradation extent, primarily because SOH serves as a gauge of a battery's general condition and ability to provide the intended performance compared to the ideal conditions [5]. As a result, precise estimation and efficient battery SOH monitoring are essential for maintaining safe, effective, and optimal energy system performance as well as for appropriate maintenance and retirement schemes [6]. Nevertheless, in practical applications, the capacity of a working battery cannot be directly measured by measuring instruments [7], indicating that estimation algorithms based on other measured signals (e.g., voltage, current, and temperature) become the only solution to acquire the value of SOH, which is a challenging task [8].

Various methods have been investigated for SOH estimation algorithms, which can be divided into three main classes as follows: experimental methods, model-based approaches, and data-driven techniques [9].

The experimental methods encompass the ampere-hour concept, the open circuit voltage approach, and impedance spectroscopy method [10]. All these approaches are capable of obtaining extensive degradation information and accurate SOH estimation outcomes straightforwardly. However, they require high-precision measurements, multiple operations, and may permanently degrade the lifespan of the battery. Hence, experimental methods are more suitable in specialized laboratories while practical applications are somewhat restricted [5].

Model-based methods generally estimate SOH by identifying the parameters of an electrochemical model or an equivalent circuit model [11]. Based on the chemical reactions and transport mechanisms occurring inside lithium-ion batteries, the electrochemical model establishes partial differential equations to describe the battery's properties [12]. It has a distinct physical meaning and offers exceptional precision, yet it demands specialized knowledge of electrochemical theory and its partial differential equations are challenging to solve [13]. Magri et al. even couple the thermal behavior of the battery with the electrochemical model, which can indeed provide a more accurate simulation of battery behavior, but this makes practical application more challenging [14]. The equivalent circuit model represents the battery as a circuit composed of resistors and capacitors[15]. Resistors characterize internal resistance and polarization internal resistance of the battery, while capacitors simulate the dynamic process of polarization phenomena. SOH estimation can be achieved by identifying the circuit component parameters combined with various Kalman filtering algorithms [16,17]. However, the parameter identification error tends to increase as the battery ages, causing the inaccuracy in SOH estimation.

Unlike model-based methods that require specialized expertise, data-driven approaches tend to establish the connection between collected operating data and SOH straightforwardly, which encompass traditional machine learning and deep learning techniques. Over the last decade, numerous machine learning algorithms have been applied to estimate SOH, ranging from k-nearest neighbor regression [18], random forest regression [19], linear regression analysis [20], and support vector machines [21] to Gaussian process regression [22] and extreme learning machine [23]. All these traditional machine learning models are able to extract complex connections between features and SOH as well as battery degradation patterns, but their accuracy and effectiveness highly depend on meticulous feature engineering.

Conversely, deep learning markedly diminishes reliance on manually engineered features, instead autonomously deriving abstract representations from raw data through multi-layered networks. This inherent capability confers a considerable advantage in the processing of complex, high-dimensional, and nonlinear data; thus, increasing attention has been directed toward exploring the employment of deep learning approaches for SOH estimation and battery prognosis [24–26]. Through leveraging deep neural networks, researchers are able to automatically extract valuable features from battery historical data, leading to more accurate and reliable SOH estimation. Convolutional neural networks

(CNNs) are proficient in gathering local variables hierarchically through their convolutional operations and preserve all local cues as feature maps [5]. For example, Yang et al. employed a CNN to extract indicators for estimating SOH [27]. Given that SOH estimation algorithms are handling sequential data to derive the final result, SOH estimation can be characterized as a time-series problem. Furthermore, recurrent neural networks like long short-term memory (LSTM) neural networks and gated recurrent unit (GRU) neural networks, are well-suited for modeling such sequences owing to their capability to capture temporal dependencies. For instance, Ma et al. applied LSTM with a differential evolution gray wolf optimizer to learn the temporal relationships for SOH estimation by using its internal memory [10]. Xu et al. leveraged the advantages of both CNN and LSTM to improve the SOH estimation accuracy [28], yielding remarkable outcomes. To further improve the monitoring of SOH, a study introduced a model that integrates CNN and bidirectional gated recurrent unit (BiGRU) networks for SOH estimation [29], which highlights that GRU offers a more streamlined architecture and better computational efficiency compared to the LSTM.

It is worth noting that whether using CNN, LSTM, or GRU, these models can only capture local or short-term features. For batteries, the entire charge-discharge process at various stages affects their health state and degradation. The Transformer model, with its multi-head self-attention mechanism, excels at global feature extraction [30]. However, it often overlooks local feature details, which can be crucial for accurate analysis. To address this constraint, the hybrid CNN-Transformer model proposed by Gu et al. combines the CNN's hierarchical local feature extraction with Transformer's global property aggregation [5], ensuring the precise estimations of battery SOH. In order to conserve the computational resources associated with attention mechanisms and optimize the efficiency of SOH estimation, Zhao et al. utilized a multi-head sparse self-attention mechanism, which proved impressive effectiveness in achieving accurate estimations and facilitating rapid training [31].

With the continuous exploration of deep learning in the field of battery SOH estimation, the CNN-Transformer hybrid model has shown excellent performance after being improved by researchers, achieving accurate SOH estimation. However, in practical applications, there are still several aspects that warrant further investigation. Firstly, the CNN used in the aforementioned models to extract local features are mostly based on the time-series property, which may not fully account for the internal connections among multiple physics measurement signals. Secondly, the Transformer attention architecture, while aggregating local features to capture global features, still faces issues like quadratic time complexity and heavy memory consumption [31]. This might impact its efficiency in certain cases. To conquer these problems, this study proposes the dimension-wise convolutions-global filter networks (DimConv-GFNet), a hybrid architecture that effectively captures both local and global dependencies across various physics measurement signals, while maintaining better computational efficiency than Transformer attention architecture.

The core innovations of DimConv-GFNet lie in two aspects as follows:

Firstly, by utilizing dimension-wise convolutions (DimConv) applied to different dimensions, the proposed model can capture inter-signal relationships at multiple scales. This allows DimConv-GFNet to handle various physical measurements (voltage, current, temperature, and time) across both temporal and spatial dimensions. By considering these signals at varying scales, the model can better represent the underlying physical behavior of the battery. This multi-scale approach improves our understanding of complex signal interactions, leading to better performance for SOH estimation in real-world applications.

Secondly, this paper employs global filter networks (GFNets) that can be regarded as a replacement for the conventional Transformer attention architectures. A GFNet operates in the Fourier frequency domain by applying the element-wise multiplication of the input spectrum with a learnable global filter. Such an operation significantly reduces computational complexity while effectively capturing global dependencies. Notably, it takes advantage of the conjugate symmetric property of the spectrum, allowing it to process only half of the spectral values. This approach not only simplifies computations but also substantially decreases overall computational costs. The lightweight global filter network enhances model efficiency, capturing long-range dependencies while remaining resource-efficient. This change is especially beneficial for real-world applications in the resource-constrained conditions, since it significantly lowers computational requirements with only a slight compromise in performance.

These advancements permit DimConv-GFNet to achieve accurate SOH estimation efficiently, making it highly suitable for battery management systems. The rest of this paper is organized as follows. Section 2 presents the architecture and design of the DimConv-GFNet model. Section 3 introduces datasets and data processing and then outlines the experimental setup. Section 4 show the experimental results and provides a discussion. Section 5 concludes the paper and discusses potential future directions.

2. Methodology

2.1. DimConv

Two-dimensional convolution can efficiently capture the local potential relationships within the input data. The two-dimensional data are processed by moving a convolutional kernel to extract the spatial latent features. Essentially, this process is a weighted summation of the data within the input region, followed by the addition of a bias term, which can be defined as follows:

$$y_{ij} = PReLU\left(\sum_{m=1}^{3}\sum_{n=1}^{3}\omega_{mn}x_{i+m-2,j+n-2} + \beta\right) \quad (1)$$

where the kernel size is 3 × 3. Therefore, in this paper, two-dimensional convolution is used to handle the mapping association between the voltage, current, temperature, time, and SOH.

DimConv (Figure 1) initially executes two-dimensional convolutions across various dimensions. Subsequently, the respective features are uniformly shuffled and fused through a further layer that computes a linear combination of the output via 1 × 1 convolution [32]. Specifically, there are different convolutional kernels sliding along the planes perpendicular to the depth, height, and width dimensions. Specifically, as the convolution kernel moves along the plane perpendicular to the depth (depth-wise convolution), it captures the features among voltage, current, temperature, and time at a smaller temporal scale, while the convolution kernel moving along the plane perpendicular to the height (height-wise convolution) learns features over a larger temporal scale. Furthermore, the convolution kernel moving along the plane perpendicular to the width dimension (width-wise convolution)extracts features representing the individual spatial characteristics of voltage, current, temperature, and time at another time scale. Then, the output of various dimensions are uniformly shuffled. A 1 × 1 convolution, with input channels that are three times that of the output channels, is ultimately used to fuse the features derived from the three dimensions with the PReLU activation function, achieving the integration of two temporal scales and the enhancement of an individual feature in this process. This results in the retrieval of multi-scale local fusion features (N, H, W), where (H, W) is the size of each patch and N is the number of patches, which also serves as the effective input sequence length for global filter network.

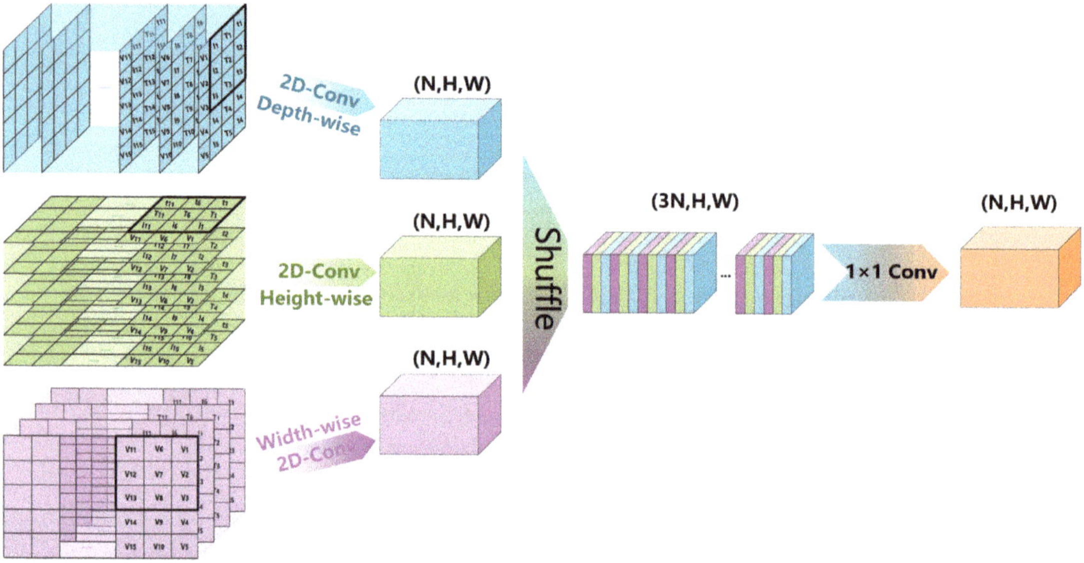

Figure 1. Dimension-wise convolutions.

2.2. GFNet

Since the input data for the global filter must be two-dimensional, a patch embedding layer is introduced to bridge the DimConv and the global filter layer. This layer flattens the local features of each patch from (H, W) to (H × W) and maps them to D dimensions using a trainable linear projection [33], finally reshaping them into (D, N).

As a fundamental technique in digital signal processing, the Fourier transform shifts signals from the time domain to the frequency domain [34]. Consequently, the global filter network is designed to facilitate global information exchange among patches in the Fourier frequency domain. This network comprises a stack of N_{depth} identical layers, each containing two sublayers. The first sublayer is a global filter layer, while the second is a feed-forward layer. Each sublayer is preceded by layer normalization to enhance model stability and mitigate issues such as gradient vanishing and explosion. Additionally, we employ a residual connection around each layer to improve gradient flow within the network [35].

2.2.1. Discrete Fourier Transform

In fact, actual data are typically discrete, thus we employ the two-dimensional discrete Fourier transform (2D DFT) to convert the local features from the temporal and spatial domain to the Fourier frequency domain. Given the 2D matrix $x[d,n]$, $0 \le d \le D-1$, $0 \le n \le N-1$, the frequency domain representation of $x[d,n]$ is given by the following:

$$X[u,v] = \sum_{d=0}^{D-1} \sum_{n=0}^{N-1} x[d,n] e^{-j2\pi\left(\frac{ud}{D} + \frac{vn}{N}\right)} \tag{2}$$

where u, v represent the frequency domain coordinates, $0 \le d \le D-1, 0 \le n \le N-1$. It is worth noting that $X[u,v]$ are conjugate symmetric, which can be proved as follows:

$$\begin{aligned} X[D-u,N-v] &= \sum_{d=0}^{D-1}\sum_{n=0}^{N-1} x[d,n]e^{-j2\pi\left[\frac{(D-u)d}{D}+\frac{(N-v)n}{N}\right]} \\ &= \sum_{d=0}^{D-1}\sum_{n=0}^{N-1} x[d,n]e^{j2\pi\left(\frac{ud}{D}+\frac{vn}{N}\right)} \\ &= X^*[u,v] \end{aligned} \tag{3}$$

where $X[D-u,N-v]$ is the complex conjugate of $X[u,v]$.

Leveraging the conjugate symmetric property, only half of the values in X need to be fed into the global filter encoder, while the full information is preserved at the same time [36]. Finally, given the 2D DFT $X[u,v]$, we can recover the original matrix $x[d,n]$ by the inverse DFT (IDFT) as follows:

$$x[d,n] = \frac{1}{DN}\sum_{u=0}^{D-1}\sum_{v=0}^{N-1} X[u,v]e^{j\,2\pi\left(\frac{ud}{D}+\frac{vn}{N}\right)} \tag{4}$$

2.2.2. Global Filter Design

As the core component of the global filter networks, the global filter needs to capture the global features among patches, whose input data must be two-dimensional. Following the flattening, mapping, and reshaping processes of the patch embedding layer, all the features are converted to the Fourier frequency domain, which can be represented by the following equation:

$$X_f = \mathcal{F}[x] \in \mathbb{C}^{D\times N} \tag{5}$$

where $\mathcal{F}[\cdot]$ denotes the 2D DFT. Note that X_f is a complex tensor and represents the spectrum of x. Since x is a real tensor and the corresponding spectrum X_f is conjugate symmetric, only half of the X_f is needed for further processing as follows [37]:

$$X_r = X_f\left[\,:, 0 : \left\lceil\frac{N}{2}\right\rceil\right] \in \mathbb{C}^{D\times\lceil\frac{N}{2}\rceil} \tag{6}$$

where $\lceil\frac{N}{2}\rceil$ is the least integer that is at least $\frac{N}{2}$. According to the spectral convolution theorem in Fourier theory, a single value modification in the spectral domain produces a global effect on the entire original data [38], providing insights into design operations with the non-local receptive field. In other words, the Fourier transform amalgamates the entire information into distinct elements of the spectrum. Consequently, we can represent the global context through a rudimentary element-wise multiplication between the spectrum X_r and a learnable global filter K as follows:

$$\tilde{X}_r = K \odot X_r \tag{7}$$

where $\odot$ is the Hadamard product, $K \in \mathbb{C}^{D\times\lceil\frac{N}{2}\rceil}$, weighting the features of each patch in the Fourier frequency domain. Ultimately, the inverse DFT (IDFT) is adopted to transform the modulated spectrum $\tilde{X}_r$ back, as follows:

$$X \leftarrow \mathcal{F}_r^{-1}\left[\tilde{X}_r\right] \tag{8}$$

Subsequent to the global filter layer, a Feed Forward Network (FFN) is utilized. The FFN, via linear transformations and nonlinear activation of neural units, substantially bolsters the feature extraction proficiency of the global filter network. FFN operations are as follows.

$$FFN(x) = GELU(xW_1 + b_1)W_2 + b_2 \tag{9}$$

2.3. Overview of DimConv-GFNet

Figure 2 shows the complete DimConv-GFNet-based SOH estimation framework. The model primarily consists of four parts, dimension-wise fusion convolutions, the global filter layer, the average pooling layer, and the multi-layer perceptron. The sensor signals of the battery (voltage, current, temperature, and time of a single cycle) are first uniformly interpolated and resampled to the required number of sample points for the model. Then, the data are divided into different patches and reshaped into a three-dimensional tensor. DimConv is applied to extract multi-scale fused local features, followed by the use of GFNet to exchange and mix information between patches to capture global features. The global features are then subjected to an average pooling layer and passed through a multi-layer perceptron to output the estimated SOH. The model workflow is summarized in Table 1.

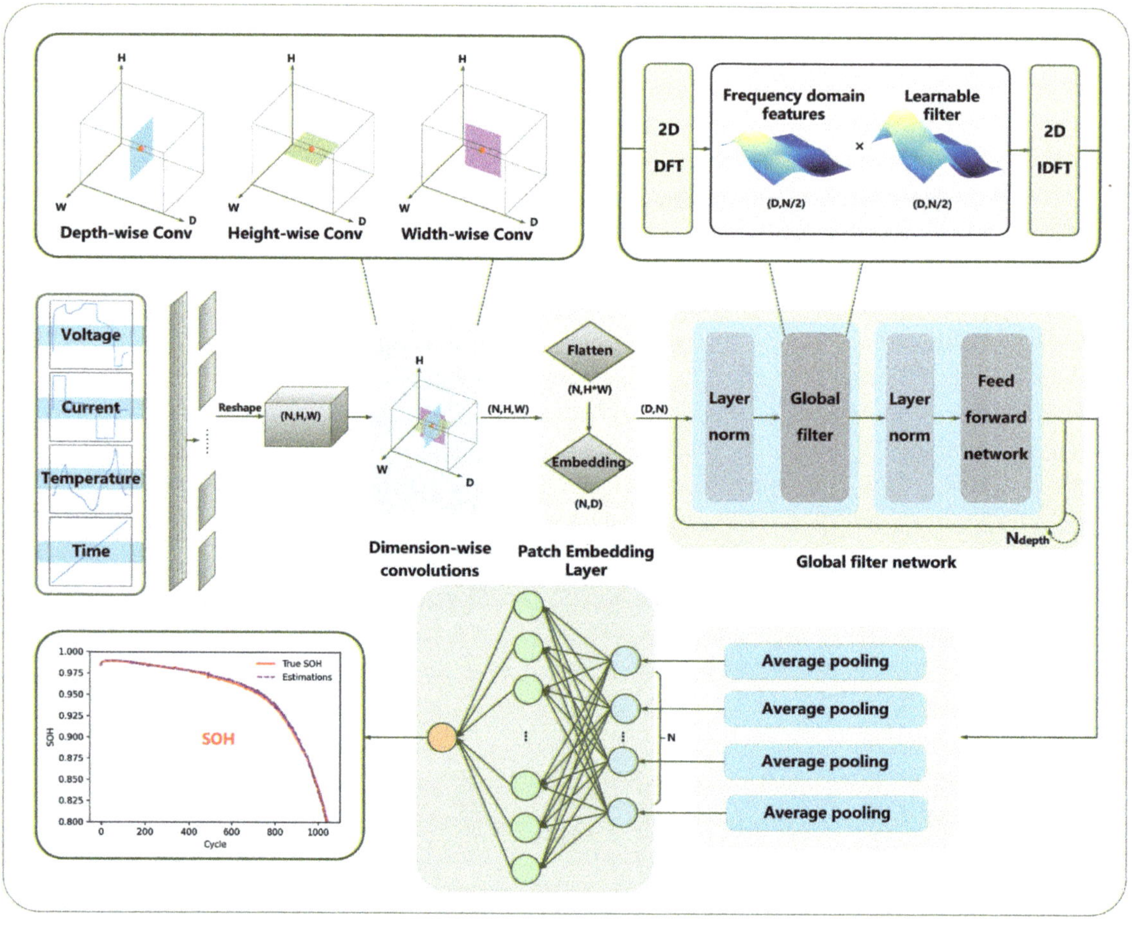

Figure 2. Framework of the proposed DimConv-GFNet for SOH estimation.

Table 1. Model workflow.

Work Step	Details
Input source data	Voltage, current, temperature, time
Data processing	Resample 480 sampling points and normalize to $[-1, 1]$
DimConv	Multi-scale fused local features
GFNet	Global weighted features
Average pooling	Compressed features
Mlp output	SOH

3. Experiment

3.1. Datasets Description

Datasets containing fast charging cycling data are originally provided by Severson et al. [39] and Attia et al. [40]. They involve a total of 169 commercial lithium ion phosphate (LFP)/graphite cells manufactured by A123 System (APR18650M1A), each with a nominal capacity of 1.1 Ah. All cells were cycled until they reached end-of-life (defined as 80% of their nominal capacity). The cell charging conditions consist of different types of multi-stage constant current (CC) charging, employing one-step, two-step, or four-step fast-charging strategies up to 80% state-of-charge (SOC) (Figure 3), mostly ranging from 3C to 8C [41]. Subsequently, the batteries were charged at 1C using a uniform constant current-constant voltage (CC-CV) protocol to a cutoff voltage of 3.6 V and a cutoff current of 0.02C. Finally, all batteries were discharged at a rate of 4C to a cutoff voltage of 2.0 V [39,40]. Voltage, current, and temperature directly from sensors were recorded during aging cycles. The discharge profiles exhibited differences across individual batteries, with most having a cycle life ranging from 500 to 1100 cycles. From Figure 3, it can be seen that the datasets provide rich voltage, current, and temperature information, which can be used by our model to further capture the underlying physical behaviors of the battery. Note that two of the 169 cells are excluded due to abnormal recorded data.

The SOH of battery is defined as the ratio of the current capacity to its nominal capacity as follows:

$$SOH = \frac{Q_{\text{current}}}{Q_{\text{nominal}}} \times 100\% \tag{10}$$

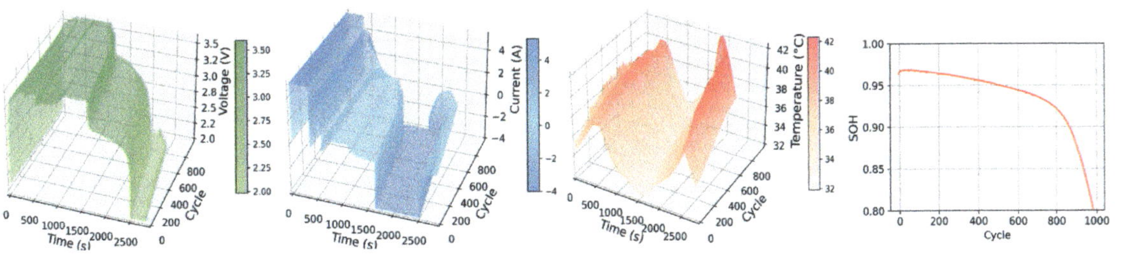

(a) (b) (c) (d)

Figure 3. Two-step fast-charging battery example b3c25. **(a)** Voltage, **(b)** current, **(c)** temperature, and **(d)** SOH.

3.2. Data Processing

Different batteries have different sampling point counts for each cycle due to their different charging policies and varying aging degrees. As shown in Figure 3, different SOH values are associated with distinct sets of voltage, current, and temperature profiles. In other words, the voltage, current, and temperature are all time-varying series that correspond to a specific SOH [42]. Nevertheless, to meet the requirement of the model, the number of sampling points must be consistent and fixed. Therefore, using time as the reference, the voltage, current and temperature data are uniformly interpolated to a fixed length of 480. To accelerate the algorithm's convergence and improve the estimation accuracy of the model, the resampled data are min–max-normalized to standardize the data scale within the range of $[-1, 1]$ [31], which is expressed as follows:

$$x_{norm} = 2 \cdot \frac{x_0 - x_{\min}}{x_{\max} - x_{\min}} - 1 \tag{11}$$

3.3. Experimental Settings

Previous studies have demonstrated that generic CNNs and GRU/LSTM models underperform individually compared to their hybrid models [5,29]. Consequently, we did not include them in the comparison with our proposed model. Instead, we compared our SOH estimation framework DimConv-GFNet with four additional models. These models included the CNN-LSTM, CNN-GRU, CNN-Transformer, and DimConv-Transformer. All the CNN components in these models consist of two layers of stacked convolutional, activation and pooling operations, while both LSTM and GRU are implemented with two layers as well. Note that the Transformer attention architecture also has a depth of two layers. Furthermore, Table 2 shows the main hyperparameters of our proposed model.

Table 2. Model hyperparameters.

Optimizer	Adam
Learning rate	0.00097
Epoch	130
Batch size	32
Dropout rate	0.119
Depth	2
Embedding dim	236
Feed forward network dim	2.5 × 236
Mlp hidden size	190

Model validation was performed on a server equipped with an AMD EPYC 7542 CPU operating at 2.9 GHz, along with an NVIDIA GeForce GTX 4090 graphics card. The proposed SOH estimation framework and the comparison methods were implemented using Python 3.8, utilizing the 'Torch' package.

4. Result and Discussion

4.1. Evaluation Criteria

In order to assess the performance of the SOH estimation methods, several evaluation metrics are utilized. These metrics include mean absolute error (MAE), mean absolute percentage error (MAPE), and root mean squared error (RMSE). RMSE quantifies the deviation between the estimated and reference values, providing an indication of the estimation accuracy. MAE reflects the average magnitude of the errors generated by the proposed method. MAPE measures the relative percentage error between the predicted and true values [43]. Formulas for the $RMSE$, MAE, and $MAPE$ are provided below.

$$RMSE = \sqrt{\frac{1}{n}\sum_{i=1}^{n}(y_i - \hat{y}_i)^2} \tag{12}$$

$$MAE = \frac{1}{n}\sum_{i=1}^{n}|y_i - \hat{y}_i| \tag{13}$$

$$MAPE = \frac{1}{n}\sum_{i=1}^{n}\left|\frac{\hat{y}_i - y_i}{y_i}\right| \quad (14)$$

4.2. Experimental Results and Discussion

Since the batteries in the two datasets are identical, except for the charging and discharging conditions, we randomly shuffled the 167 batteries from both datasets. Then, we chose 117 batteries (70%) for training the model, and the remaining 50 batteries (30%) were used to evaluate the performance of the proposed method.

Figure 4 and Table 3 illustrate the overall test results. The CNN-Transformer, DimConv-Transformer, and DimConv-GFNet models demonstrate superior SOH estimation accuracy compared to the CNN-GRU and CNN-LSTM frameworks, which focus solely on time-series characteristics. The RMSE, MAE, and MAPE for the former models are all lower than those of the latter. According to Table 3, DimConv-Transformer achieved a RMSE of 0.227%, MAE of 0.146%, and MAPE of 0.144%, compared to CNN-Transformer's 0.365% RMSE, 0.242% MAE, and 0.240% MAPE. This demonstrates that dimension-wise convolutions contribute to improved SOH estimation accuracy. In Figure 4, the box plot for DimConv-Transformer shows a significantly reduced height, narrower spread, and a more concentrated error range, further indicating that by adequately learning the underlying relationships among battery operating signals, we can ensure the stability of SOH estimation accuracy to a certain extent.

Table 3. Comparison of estimation results.

Model	RMSE (%)	MAE (%)	MAPE (%)
CNN-GRU	0.632	0.479	0.472
CNN-LSTM	0.612	0.462	0.454
CNN-Transformer	0.365	0.242	0.240
DimConv-Transformer	0.227	0.146	0.144
DimConv-GFNet (ours)	0.335	0.233	0.230

To evaluate the complexity and efficiency of our deep learning model for SOH estimation, we consider the following two key metrics: Floating Point Operations (FLOPs) and Parameters (Params). FLOPs quantify the sum number of floating-point operations required for a single forward pass of the model. Generally speaking, a higher FLOP counts typically suggests a more complicated model that takes more processing resources [44]. Furthermore, Params represent the number of trainable parameters within the model, directly impacting the model's storage needs and training complexity. Although DimConv-GFNet exhibits a RMSE of 0.335%, MAE of 0.233%, and MAPE of 0.230%, which are slightly inferior in accuracy compared to DimConv-Transformer, Table 4 indicates that the FLOPs and parameters of DimConv-GFNet are less than one-third of those of DimConv-Transformer, with both models having the same depth of two layers. The FLOPs and Params of DimConv-GFNet are less than one-third of those of DimConv-Transformer, demonstrating that the GFNet in this paper maintains the ability to learn global information while also being efficient and having lower computational complexity than the Transformer.

Table 4. Comparison of model complexity.

Model	FLOPs	Params
DimConv-Transformer	208703559	2093737
DimConv-GFNet (ours)	54141494	656048

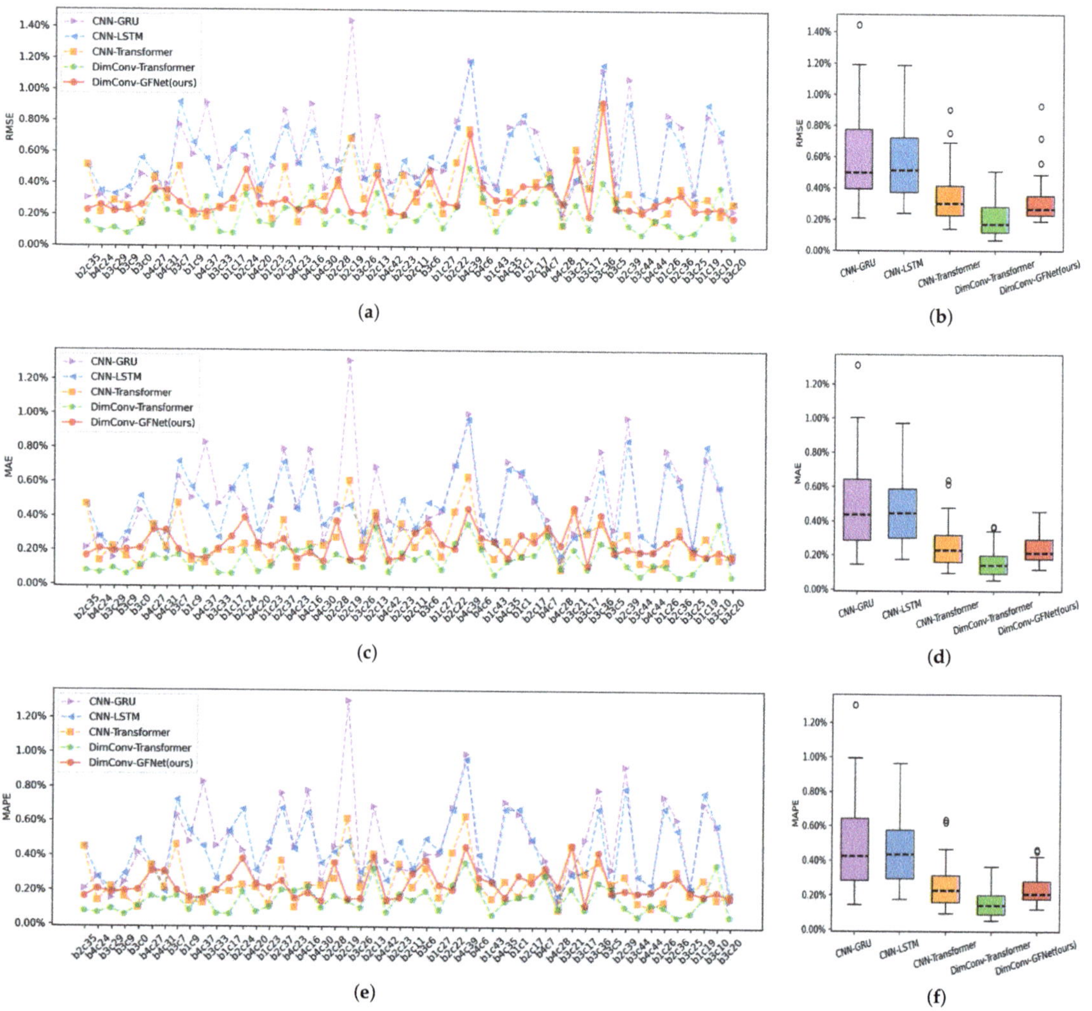

Figure 4. Comparison of different models for all test batteries. (**a**,**b**) RMSE, (**c**,**d**) MAE, and (**e**,**f**) MAPE.

Figure 5 shows some SOH estimation results for batteries from four batches. It can be observed that the estimation results of DimConv-GFNet are not only closer to the true values in terms of overall trends but also more accurate at points of capacity fluctuation. The absolute error plots on the right side of Figure 5 clearly demonstrate that the DimConv-GFNet model (red line) consistently exhibits the lower estimation errors across all battery cycles. In contrast, other models experience significant error increases, particularly in the later stages. This indicates that the DimConv-GFNet model offers considerable efficiency and accuracy in SOH estimation tasks.

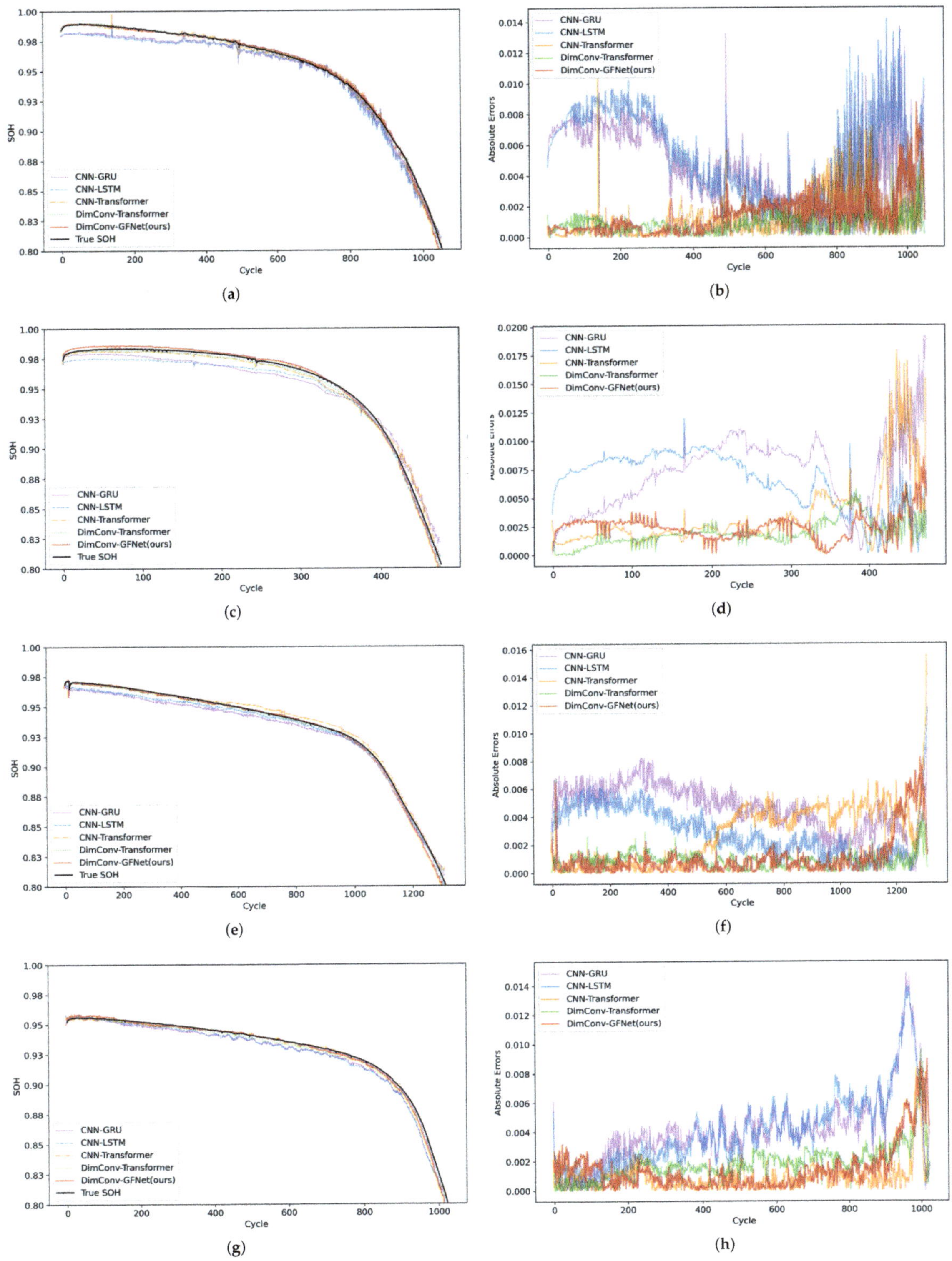

Figure 5. (**a**,**b**) b1c9 results (**c**,**d**) b2c37 results (**e**,**f**) b3c17 results (**g**,**h**) b4c23 results.

5. Conclusions

An innovative hybrid DimConv-GFNet model for SOH estimation is proposed in this paper, which is formed using a combination of dimension-wise convolutions and a global filter net. By exploring the relationships among voltage, current, temperature, and time, DimConv successfully extracts multi-scale local features. The following GFNet further captures global long-range dependencies in the Fourier frequency domain efficiently. Compared to traditional Transformer attention architectures, the GFNet shows improved performance regarding complexity and computational efficiency, achieving precise and efficient SOH estimation.

Although this study demonstrates superior accuracy and computational efficiency with DimConv-GFNet, there still remain some limitations. For example, the model requires fixed-length data for the entire cycle, which is a problem to be improved for some special battery operational conditions in the real word. Compared to traditional machine learning methods, DimConv-GFNet still needs to be further simplified. In addition, the selection of sampling points and the division of patches require further exploration. Future research will aim to improve the aforementioned limitations, thereby strengthening its applicability to real-world operating batteries.

To conclude, the introduced DimConv-GFNet model represents an important step forward in the field of battery SOH estimation, offering a novel perspective for deep learning-based battery state estimation algorithms. Ongoing refinements and optimizations in the future may facilitate its application in engineering practice.

Author Contributions: Conceptualization, K.H.; methodology, K.H.; validation, K.H., J.K., and O.W.; formal analysis, K.H., J.K., and O.W.; investigation, K.H.; resources, K.H.; data curation, K.H.; writing—original draft preparation, K.H.; writing—review and editing, K.H. and O.W.; visualization, K.H. and O.W.; supervision, J.K., J.V.W., Q.W., and O.W.; funding acquisition, J.K. All authors have read and agreed to the published version of the manuscript.

Funding: This research was funded by the National Natural Science Foundation of China grant number 52277224 and National Key Laboratory of Electromagnetic Energy under grant number 61422172320104.

Data Availability Statement: The data that support the findings of this study are openly available at URL/DOI: https://data.matr.io/1, accessed on 21 August 2024.

Acknowledgments: The authors would like to thank Severson and Attia for providing the original battery cycling data.

Conflicts of Interest: The authors declare no conflict of interest.

Abbreviations

The following abbreviations are used in this manuscript:

SOH	State of health
DimConv	Dimension-wise convolutions
GFNet	Global filter network
FLOPs	Floating point operations
Params	Parameters

References

1. Vignesh, S.; Che, H.S.; Selvaraj, J.; Tey, K.S.; Lee, J.W.; Shareef, H.; Errouissi, R. State of Health (SoH) estimation methods for second life lithium-ion battery—Review and challenges. *Appl. Energy* **2024**, *369*, 123542.
2. Masias, A.; Marcicki, J.; Paxton, W.A. Opportunities and challenges of lithium ion batteries in automotive applications. *ACS Energy Lett.* **2021**, *6*, 621–630. [CrossRef]

3. Costa, C.M.; Barbosa, J.C.; Gonçalves, R.; Castro, H.; Del Campo, F.; Lanceros-Méndez, S. Recycling and environmental issues of lithium-ion batteries: Advances, challenges and opportunities. *Energy Storage Mater.* **2021**, *37*, 433–465. [CrossRef]
4. Yang, S., Zhang, C., Jiang, J., Zhang, W., Zhang, L., Wang, Y. Review on state-of-health of lithium-ion batteries: Characterizations, estimations and applications. *J. Clean. Prod.* **2021**, *314*, 128015. [CrossRef]
5. Gu, X.; See, K.W.; Li, P.; Shan, K.; Wang, Y.; Zhao, L.; Lim, K.C.; Zhang, N. A novel state-of-health estimation for the lithium-ion battery using a convolutional neural network and transformer model. *Energy* **2023**, *262*, 125501. [CrossRef]
6. Cao, Z.; Gao, W.; Fu, Y.; Kurdkandi, N.V.; Mi, C. A general framework for lithium-ion battery state of health estimation: From laboratory tests to machine learning with transferability across domains. *Appl. Energy* **2025**, *262*, 125086. [CrossRef]
7. Chen, L.; Xie, S.; Lopes, A.M.; Li, H.; Bao, X.; Zhang, C.; Li, P. A new SOH estimation method for Lithium-ion batteries based on model-data-fusion. *Energy* **2024**, *286*, 129597. [CrossRef]
8. Ding, P.; Liu, X.; Li, H.; Huang, Z.; Zhang, K.; Shao, L.; Abedinia, O. Useful life prediction based on wavelet packet decomposition and two-dimensional convolutional neural network for lithium-ion batteries. *Renew. Sustain. Energy Rev.* **2021**, *148*, 111287. [CrossRef]
9. Demirci, O.; Taskin, S.; Schaltz, E.; Demirci, B.A. Review of battery state estimation methods for electric vehicles-Part II: SOH estimation. *J. Energy Storage* **2024**, *96*, 112703. [CrossRef]
10. Ma, Y.; Shan, C.; Gao, J.; Chen, H. A novel method for state of health estimation of lithium-ion batteries based on improved LSTM and health indicators extraction. *Energy* **2022**, *251*, 123973. [CrossRef]
11. Fu, S.; Tao, S.; Fan, H.; He, K.; Liu, X.; Tao, Y.; Zuo, J.; Zhang, X.; Wang, Y.; Sun, Y. Data-driven capacity estimation for lithium-ion batteries with feature matching based transfer learning method. *Appl. Energy* **2024**, *353*, 121991. [CrossRef]
12. Che, Y.; Hu, X.; Lin, X.; Guo, J.; Teodorescu, R. Health prognostics for lithium-ion batteries: Mechanisms, methods, and prospects. *Energy Environ. Sci.* **2023**, *16*, 338–371. [CrossRef]
13. Chen, S.-Z.; Liu, J.; Yuan, H.; Tao, Y.; Xu, F.; Yang, L. AM-MFF: A multi-feature fusion framework based on attention mechanism for robust and interpretable lithium-ion battery state of health estimation. *Appl. Energy* **2025**, *381*, 125116. [CrossRef]
14. Magri, L.; Sequino, L.; Ferrari, C. Simulating the Electrochemical-Thermal Behavior of a Prismatic Lithium-Ion Battery on the Market under Various Discharge Cycles. *Batteries* **2023**, *9*, 397. [CrossRef]
15. Tran, M.-K.; Mathew, M.; Janhunen, S.; Panchal, S.; Raahemifar, K.; Fraser, R.; Fowler, M. A comprehensive equivalent circuit model for lithium-ion batteries, incorporating the effects of state of health, state of charge, and temperature on model parameters. *J. Energy Storage* **2021**, *16*, 103252. [CrossRef]
16. Yi, B.; Du, X.; Zhang, J.; Wu, X.; Hu, Q.; Jiang, W.; Hu, X.; Song, Z. Bias-compensated state of charge and state of health joint estimation for lithium iron phosphate batteries. *IEEE Trans. Power Electron.* **2024**, *40*, 3033 - 3042. [CrossRef]
17. Mu, A.X.; Zhang, B.J.; Li, C.G.; Xiao, D.Z.; Zeng, E.F.; Liu, F.J. Estimating SOC and SOH of energy storage battery pack based on voltage inconsistency using reference-difference model and dual extended Kalman filter. *J. Energy Storage* **2024**, *81*, 110221. [CrossRef]
18. Hu, C.; Jain, G.; Zhang, P.; Schmidt, C.; Gomadam, P.; Gorka, T. Data-driven method based on particle swarm optimization and k-nearest neighbor regression for estimating capacity of lithium-ion battery. *Appl. Energy* **2014**, *129*, 49–55. [CrossRef]
19. Li, Y.; Zou, C.; Berecibar, M.; Nanini-Maury, E.; Chan, J.C.-W.; Van den Bossche, P.; Van Mierlo, J.; Omar, N. Random forest regression for online capacity estimation of lithium-ion batteries. *Appl. Energy* **2018**, *232*, 197–210. [CrossRef]
20. Kang, S.-H.; Noh, T.-W.; Lee, B.-K. Machine learning-based SOH estimation algorithm using a linear regression analysis. *Trans. Korean Inst. Power Electron.* **2021**, *26*, 241–248.
21. Zhang, Y.; Liu, Y.; Wang, J.; Zhang, T. State-of-health estimation for lithium-ion batteries by combining model-based incremental capacity analysis with support vector regression. *Energy* **2022**, *239*, 121986. [CrossRef]
22. Cai, L.; Lin, J. A charging-feature-based estimation model for state of health of lithium-ion batteries. *Expert Syst. Appl.* **2024**, *238*, 122034. [CrossRef]
23. Sui, X.; He, S.; Teodorescu, R. Small-sample-learning-based lithium-ion batteries health assessment: An optimized ensemble framework. *IEEE Trans. Ind. Appl.* **2024**, *60*, 4366–4380. [CrossRef]
24. Zhao, J.; Han, X.; Ouyang, M.; Burke, A.F. Specialized deep neural networks for battery health prognostics: Opportunities and challenges. *J. Energy Chem.* **2023**, *87*, 416–438. [CrossRef]
25. Yuan, Z.; Tian, T.; Hao, F.; Li, G.; Tang, R.; Liu, X. A hybrid neural network based on variational mode decomposition denoising for predicting state-of-health of lithium-ion batteries. *J. Power Sources* **2024**, *609*, 234697. [CrossRef]
26. Guo, R.; Zhang, K.; He, S.; Tao, S.; Zhang, X.; Liu, K.; Li, X.; Tian, J.; Shen, W.; Chung, C.Y. Robust Health Monitoring for Lithium-Ion Batteries under Guidance of Proxy Labels: A Deep Multi-Task Learning Approach. *IEEE Trans. Power Electron.* **2025**. [CrossRef]
27. Yang, N.; Song, Z.; Hofmann, H.; Sun, J. Robust State of Health estimation of lithium-ion batteries using convolutional neural network and random forest. *J. Energy Storage* **2022**, *48*, 103857. [CrossRef]

28. Xu, H.; Wu, L.; Xiong, S.; Li, W.; Garg, A.; Gao, L. An improved CNN-LSTM model-based state-of-health estimation approach for lithium-ion batteries. *Energy* **2023**, *276*, 127585. [CrossRef]
29. Mazzi, Y.; Sassi, H.B.; Errahimi, F. Lithium-ion battery state of health estimation using a hybrid model based on a convolutional neural network and bidirectional gated recurrent unit. *Eng. Appl. Artif. Intell.* **2024**, *127*, 107199. [CrossRef]
30. Zhou, H.; Zhang, Y.; Yang, L.; Liu, Q.; Yan, K.; Du, Y. Short-term photovoltaic power forecasting based on long short term memory neural network and attention mechanism. *IEEE Access* **2019**, *7*, 78063–78074. [CrossRef]
31. Zhao, J.; Wang, Z. Specialized convolutional transformer networks for estimating battery health via transfer learning. *Energy Storage Mater.* **2024**, *71*, 103668. [CrossRef]
32. Howard, A.G.; Zhu, M.; Chen, B.; Kalenichenko, D.; Wang, W.; Weyand, T.; Andreetto, M.; Adam, H. Mobilenets: Efficient convolutional neural networks for mobile vision applications. *arXiv* **2017**, arXiv:1704.04861.
33. Dosovitskiy, A.; Beyer, L.; Kolesnikov, A.; Weissenborn, D.; Zhai, X.; Unterthiner, T.; Dehghani, M.; Minderer, M.; Heigold, G.; Gelly, S. An image is worth 16x16 words: Transformers for image recognition at scale. *arXiv* **2020**, arXiv:2010.11929
34. Kollar, Z.; Plesznik, F.; Trumpf, S. Observer-based recursive sliding discrete Fourier transform [tips & tricks]. *IEEE Signal Process. Mag.* **2018**, *35*, 100–106.
35. Vaswani, A.; Shazeer, N.; Parmar, N.; Uszkoreit, J.; Jones, L.; Gomez, A.N.; Kaiser, Ł.; Polosukhin, I. Attention is all you need. *arXiv* **2017**, arXiv:1706.03762.
36. Liu, Y.; Yu, R.; Wang, J.; Zhao, X.; Wang, Y.; Tang, Y.; Yang, Y. Global spectral filter memory network for video object segmentation. In Proceedings of the European Conference on Computer Vision, Tel Aviv, Israel, 23–27 October 2022; pp. 648–665.
37. Rao, Y.; Zhao, W.; Zhu, Z.; Lu, J.; Zhou, J. Global filter networks for image classification. *Adv. Neural Inf. Process. Syst.* **2021**, *34*, 980–993.
38. Li, Y.; Gan, J.; Lin, X.; Qiu, Y.; Zhan, H.; Tian, H. DS-TDNN: Dual-stream time-delay neural network with global-aware filter for speaker verification. *IEEE/ACM Trans. Audio, Speech, Lang. Process.* **2024**, *32*, 2814–2827. [CrossRef]
39. Severson, K.A.; Attia, P.M.; Jin, N.; Perkins, N.; Jiang, B.; Yang, Z.; Chen, M.H.; Aykol, M.; Herring, P.K.; Fraggedakis, D. Data-driven prediction of battery cycle life before capacity degradation. *Nat. Energy* **2019**, *4*, 383–391. [CrossRef]
40. Attia, P.M.; Grover, A.; Jin, N.; Severson, K.A.; Markov, T.M.; Liao, Y.-H.; Chen, M.H.; Cheong, B.; Perkins, N.; Yang, Z. Closed-loop optimization of fast-charging protocols for batteries with machine learning. *Nature* **2020**, *578*, 397–402. [CrossRef]
41. Kim, S.; Jung, H.; Lee, M.; Choi, Y.Y.; Choi, J.-I. Model-free reconstruction of capacity degradation trajectory of lithium-ion batteries using early cycle data. *eTransportation* **2023**, *17*, 100243. [CrossRef]
42. Zhu, C.; Gao, M.; He, Z.; Wu, H.; Sun, C.; Zhang, Z.; Bao, Z. State of health prediction for li-ion batteries with end-to-end deep learning. *J. Energy Storage* **2023**, *65*, 107218. [CrossRef]
43. Wang, S.; Takyi-Aninakwa, P.; Jin, S.; Yu, C.; Fernandez, C.; Stroe, D.-I. An improved feedforward-long short-term memory modeling method for the whole-life-cycle state of charge prediction of lithium-ion batteries considering current-voltage-temperature variation.*Energy* **2022**, *254*, 124224. [CrossRef]
44. Chen, Y.; Huang, X.; He, Y.; Zhang, S.; Cai, Y. Edge–cloud collaborative estimation lithium-ion battery SOH based on MEWOA-VMD and Transformer. *J. Energy Storage* **2024**, *99*, 113388. [CrossRef]

MDPI AG
Grosspeteranlage 5
4052 Basel
Switzerland
Tel.: +41 61 683 77 34

Batteries Editorial Office
E-mail: batteries@mdpi.com
www.mdpi.com/journal/batteries

www.ingramcontent.com/pod-product-compliance
Lightning Source LLC
Chambersburg PA
CBHW061603210926
51850CB00009B/2201
9783725861873